A catalogue record for this book is available from the British Library

ISBN 978-0-9926650-1-2

Published by Cuesta Ltd.

Acknowledgements

Figure 5.8 reproduced with permission from S.J. Kline, W.C. Reynolds, F.A. Schraub and P.W. Runstadler, 'The structure of turbulent boundary layers', Journal of Fluid Mechanics, Volume 30(4), pp 741-773, (1967).

Introduction

The rationale for this book is as reference material for taught courses in fluid mechanics and thermodynamics for engineering undergraduates in years 1 and 2. An incremental approach is used, where basic conceptual ideas are first introduced in a simple way suitable for engineering undergraduates starting out on their degree. Then these *same* ideas are developed more generally and completely, in a way typical of year 2 of an undergraduate engineering degree. The student is then ready for specialised courses, supported by specific texts, in the final years of the specific engineering theme being taken.

Why *Another* ThermoFluids textbook?

I have thought long and hard before deciding to write an undergraduate thermofluids text. I would rather not, and instead use something already available, but I have yet to find a book that takes a typical engineering student from the start of year 1 to the start of year 3 in one go and evolves that understanding along the way. The students seem to be forced to buy several books pitched at different levels to cover core material and this just seems unfair. This book also does not seek to emulate the standard texts from the major publishers, which include lots of colour, examples, usually a vast array of web resources, DVDs and so on. I take the view that the lecturers who delivers your undergraduate thermofluids courses know their stuff and provide you with lecture slides which they explain, example questions and other questions for you to try yourself.

Therefore the objective of this book is to provide a precise, concise and clear incremental evolution of the basic physical concepts and core material for fluid mechanics and thermodynamics for first and second year engineering undergraduate students without any messing about. The challenge of developing a new introductory 'thermofluids' course, and the dearth of *well priced* and *appropriate* textbooks on the subject inspired me to write my own. I also saw no reason to give the rights to a publisher when none of the material is new and self-publishing is so straightforward. Taking this route allows me to keep the cost down to $< 25\%$ of the combined cost of the alternatives, some of which are listed in appendix A on page 160.

Organisation of the Material in this Book

Thermofluids is initially a difficult subject on a conceptual level, so this book introduces you to the subject in five stages.

- Chapter 1 and 2 introduce all the key mathematical and conceptual building blocks of the subject. Everything from chapter 3 onwards is built on these basic laws and principles, many of which you are already probably familiar.
- Chapters 3 to 5 formally develop the subject at an introductory level. During these chapters concepts are introduced with some mathematical and physical simplification, the idea is to get the message across without getting too bogged down in equations. Chapter 3 considers energy conservation for a fixed mass of fluid and focuses on the efficiency of energy conversion processes and the impact of entropy change. Chapter 4 analyses mass, momentum and energy conservation of an inviscid fluid over a fixed volume and chapter 5 provides an introduction to viscous fluid flow and boundary layers.
- Chapter 6 marks the point at which the mathematical language becomes more complete and powerful and in chapter 8 the full form of the governing equations are defined and non-dimensional numbers take on their full meaning.

- Chapter 9 marks the point at which the material becomes more specialised and from here each chapter may be regarded as an independent subject and read in any order. Chapters 9 to 14 cover most of the material most engineering students need for most of year 2 of the thermofluids content of their degree. At present these chapters cover potential flow, compressible flow, non-ideal thermodynamics, heat transfer, turbulence and computational fluid dynamics.

Extensive use of referencing of other sections, with page numbers, is used to link backwards (to remind the reader where an assumed item of information was previously defined when developing a new idea) and also forwards (so that the reader knows why they need to understand what they are reading). This, with the very detailed table of contents obviates the need for an index, and hopefully makes the book easier to read.

Who is this book for (student version)

If you are a first or second year undergraduate student of mechanical, chemical, aeronautical, marine or civil engineering this book is for you. Also this book is a suitable (and cheap) text for other science degrees where core knowledge of fluid mechanics and thermodynamics is required, for instance environmental science and meteorology. It is designed to support the lectures and examples you are given and help you answer the questions you are going to try to solve. It may also help you if you are taking courses online.

Who is this book for (instructors version)

This book delivers the majority of core thermodynamics and fluid dynamics content for the majority of engineering degrees over years one and two, without colour or lots of examples and without missing out on any of the necessary mathematical framework. It does not skip much, but there is not much padding. It delivers the material incrementally, in more-or-less the order the students are actually taught the material over years 1 and 2, which sadly seems somewhat revolutionary in the field of thermofluids textbooks. It is designed to be part of a taught course, as reference material, so you just have sort out the overheads and provide the examples relevant to your engineering theme.

About the author

I studied Chemical Engineering as a first degree, studied a mechanical engineering based PhD and have taught fluid mechanics and thermodynamics to first, second and third year undergraduates at Imperial College London and the University of Southampton, both top rated UK Universities. The second and third year courses were given to mechanical engineering students, whilst the first year course has been delivered to a cohort of ~ 250 mechanical, aeronautical and ship science students. Currently the combined fluid mechanics and thermodynamics content is given to a cohort of ~ 500 which now includes civil engineering students and students from the Institute of Sound and Vibration Research. In addition, I have taught fluid mechanics (turbulence) at graduate level at the University of Illinois at Chicago and aerodynamics to students at the University of Canterbury, New Zealand. You can find more out about me at www.pnume.org.

Table of Contents

List of Tables

1 | Reference Material

1.1 Glossary

Certain words in thermofluids have a very specific meaning and you need to recognise the words and the meaning accurately to understand the subject. Below is an alphabetical list of important words, a brief explanation and a page reference where they are used in the text so you can see the word in action. Up until the end of chapter 5 these words are in certain places <u>underlined</u> for added emphasis to make sure you notice them and understand their significance. Beyond chapter 5 this glossary is no longer added to, in the spirit of trying to keep the early information load as small as possible.

Word	Meaning	page
Adiabatic (process)	A process without heat transfer (isentropic if reversible)	39
Centre of ...	... Area, pressure.. A point through which the integral effect can be said to act	13
Closed process	A process operating on a fixed mass (same molecules) of system	14
Control volume	A fixed volume through which fluid may flow and forces may act. Often abbreviated CV	14
Convective transport	Information transported by the flow field	23
Diffusive transport	Information transported by molecular collisions	23
Environment	The area outside the system boundary	16
Eulerian (variable)	A variable defined with respect to a fixed point in space	8
Extensive (variable)	Quantity for a fixed mass (see intensive)	15
Flow (rate)	Quantity per unit time	17
Flux	Flow (through a surface) per unit area	23
Free stream (velocity)	The velocity of the fluid far away from regions of interest	26
Form Drag	Drag due to pressure drop, turbulence and separation	85
Friction Drag	Drag due to viscous forces	84
Fully Developed	A flow with nothing changing (except maybe the pressure) in the direction of flow	25
Gradient	Spatial change in a variable, drives diffusive transport	89
Ideal (gas)	A gas which obeys equation 2.17 on page 32	32
Intensive (variable)	Quantity on a per unit (mass) basis (see extensive)	15
Internal equilibrium	No changes or gradients in the volume concerned	36
Isentropic (process)	Constant entropy process	41
Isochoric (process)	Constant volume process	38
Isobaric (process)	Constant pressure process	39
Isothermal (process)	Constant temperature process	39
Inviscid (fluid)	Without viscosity	58
Lagrangian (variable)	A variable that moves with an object relative to a stationary observer	14

Newtonian (fluid)	A fluid with a linear stress-strain relationship	21
Normal (force)	A force applied parallel to the normal to that area	21
Open (process)	A process operation involving flow through a (control) volume surface	58
Orthogonal	At right angles to	8
Origin	The zero position of an Eulerian Coordinate system.	8
Partial Derivative	Part of a gradient in more than one direction	11
Path	The route taken by the process to change a state	37
Perfect (Gas)	A gas which has constant specific heats	32
Potential core	A region of fluid uninfluenced by boundary layers	83
Process	A method to change the state of a fluid	37
Quasi-equilibrium	A change between one equilibrium state and another such that internal equilibrium is maintained during the change	37
Reversible (process)	A process able to go from A to B and B to A. Heat transfer for example is not reversible unless over zero temperature change	37
Shear (force)	A force applied tangential to that area	22
Stagnation (variable)	The value of the variable in a stationary fluid. Sometimes called total $\sim$	128
State	A fluid in a defined type of internal equilibrium	23
Steady	Not changing in time	25
System	A fixed mass of fluid that undergoes work and energy transfers through the surface	17
Thermal reservoir	Part of the environment. A thermal mass so large that however much thermal energy donated or removed the temperature does not change.	45
Total derivative	Change in a variable due to change in one other variable only	11
Transport	Moving fluid information from one spatial location to another by convection or diffusion	23
Two property rule	All information about a gas is defined by two properties at that state	36
Uniform (flow)	Constant velocity flow	25

1.2   Coordinate Systems

Two coordinate systems are used in this book. They are defined below and explained in section 2.1 on page 8.

Symbol	Units	Quantity	Comments	page
(x,y,z)	m	Position	Cartesian Coordinate System, sometimes (x_1,x_2,x_3)	8
(u,v,w)	ms^{-1}	Velocity components in (x,y,z) directions	Sometimes (u_1,u_2,u_3) or (u_x,u_y,u_z)	8

| (r,θ,z) | m | Radial, circumferential, axial position | Cylindrical Coordinate System | 8 |
| (u_r,u_θ,u_z) | ms^{-1} | Radial, circumferential, axial velocity | Cylindrical Coordinate System | 8 |

1.3 Basic Vector Notation

Here the information is relevant to the end of chapter 5 on page 75, which is designed to get the conceptual message across without too much of the mathematical baggage. This information you need is given below and it is explained in section 2.2.1 on page 9. Chapter 6 introduces vector and tensor notation fully and the remainder of the book uses that knowledge.

Symbol	Units	Quantity	Comments	page
φ	$[\varphi]$	Scalar quantity	Magnitude – one number – pressure temperature	9
$\vec{\varphi}$	$[\varphi]$	Vector Quantity	Magnitude and direction – three numbers, one for each coordinate direction see (u,v,w) above	9
$(\vec{\varphi}\cdot\vec{\varphi})$	$[\varphi]^2$	Scalar dot Product of two vectors	Defines vector magnitude and components in desired directions	10

1.4 Nomenclature : English Symbols

All of the English symbols used to the end of chapter 5 on page 75 are given here, along with a page number where they are used in anger. Comments are provided to reduce confusion. Ditto for the Greek, subscript, superscript and other math nomenclature used throughout the book. Note the way in which lower case symbols are used to represent *specific* (i.e. per unit mass) forms of upper case symbols for key variables, for instance energy.

Symbol	Units	Quantity	Comments	page
a	ms^{-1}	Speed of sound	speed of a pressure wave $a = (\gamma RT)^{1/2}$	128
$\vec{a}$	ms^{-2}	Acceleration	for example, gravity	8
A	m^2	Area	Can be a vector (area normal defined)	13
COP	–	Coefficient of Performance	"efficiency" for heat pumps and refrigerators, > 1, so not called "efficiency"	47
C_p	$Jkg^{-1}K^{-1}$	Specific heat at constant pressure	A quantity of heat required to raise the temperature of a fluid at constant pressure, which for a compressible gas includes displacement work – related to enthalpy	32

C_v	$Jkg^{-1}K^{-1}$	Specific heat at constant volume	A quantity of heat required to raise the temperature of a fluid at constant volume – related to internal energy	32
d	m	(pipe) diameter		77
D_{ab}	m^2s^{-1}	Species Diffusion Coefficient	of species a in b	24
e, E	Jkg^{-1}, J	Specific energy, Energy	e_u – internal energy (C_vT) in many textbooks u, e_h – enthalpy (e_u+p/ρ) in many textbooks h, e_k = kinetic energy ($u^2/2$) in many textbooks k, e_g = potential energy (g_z), e_p - pressure energy [flow systems only] (p/ρ)	15
g	ms^{-2}	Acceleration due to gravity	strictly a vector $\vec{g}$	68
H	–	Boundary Layer Shape Factor		85
k	$Wm^{-2}K^{-1}$	Thermal conductivity	Thermal energy diffusion coefficient	24
k_B	JK^{-1}	Boltzmann constant	relates molecular thermal energy to temperature	31
K	Nm^{-2}	Bulk Modulus		22
m	kg	Mass		15
m_a	–	Mass fraction (concentration) of species a	mass of dye per mass of water	24
m_u	$kgkmol^{-1}$	Molecular mass	The mass of 1 kmol of molecules.	31
M_x	N	Momentum in the x-direction	strictly a vector, $\vec{M}$. u_x is specific momentum	15
n	mol	Number of moles		31
N	-	Number of molecules		31
N_A	mol^{-1}	Avagadro's Number	$\sim 6\cdot10^{23}mol^{-1}$, the number of molecules in one mole	31
p	Nm^{-2}	Pressure	It is a stress magnitude, not a force. "Pascals" equivalent to Nm^{-2}. "Bar" is 10^5 Pa, 1 atm ~ 1.025bar. Several meanings, see note on page 61	68
p_v	Nm^{-2}	Vapour pressure	A fluid property, a liquid boils when this becomes equal to atmospheric pressure	61
P	W, Js^{-1}	Power	Rate of energy change	8
q, Q	Jkg^{-1}, J	Specific thermal heat, heat	Intensive/extensive forms.	15
$\dot{q}_x$	Wm^{-2}	Heat flux in the x-direction	a vector $\vec{q}$	23
R_u	$Jmol^{-1}K^{-1}$	Universal gas constant	~ 8.314 for all gases	31

R	$Jkg^{-1}K^{-1}$	Mass gas constant	287 for air	31
s,S	$Jkg^{-1}K^{-1}$, JK^{-1}	Entropy		39
T	K	Temperature	Other units possible	31
T_θ	Nm	Torque	Rotational Moment	70
v,V	m^3kg^{-1}, m^3	Specific volume, volume	Specific volume is $1/\rho$	33
w,W	Jkg^{-1}, J	specific work, work transfer		15

1.5   Nomenclature : Greek Symbols

Symbol	Units	Quantity	Comments	page
δ	m	Boundary layer thickness		85
δ^*	m	Displacement thickness		85
η	-	Efficiency	what you get in terms of what you pay for...	46
ρ	kgm^{-3}	Density		12
σ	Nm^{-2}	Normal stress	Force is acting in the same direction as the normal to a surface	107
τ	Nm^{-2}	Shear Stress	Force acting tangential to the surface	107
γ	−	Ratio of specific heats	C_p/C_v	33
$\dot\theta$	$rads^{-1}$	Rotational speed (often called angular speed and written as ω)		70
$\vec\omega$	s^{-1}	Vorticity vector		108
μ	Nsm^{-2}	Dynamic viscosity	Momentum diffusion coefficient	24
ν	ms^{-2}	Kinematic viscosity	$\nu = \mu/\rho$	24
θ	m	Momentum boundary layer thickness		85

1.6   Nomenclature : Subscripts

Symbol	Units	Quantity	Comments	page
φ_{12}	$[\varphi]$	Process transfer of φ from state 1 to state 2	Units depend on quantity transferred	51
φ_o	$[\varphi]$	Reference, dimensional, value of φ	Used with φ^*, e.g. $\varphi = \varphi_o\varphi^*$	29

1.7   Nomenclature : Superscripts

Symbol	Units	Quantity	Comments	page
$\dot{Q}$	Js^{-1}	Rate of change of Q with time	Always time, can be V, m	12
$\overline{\varphi}$	$[\varphi]$	Average value of φ		77
φ^*	$[\varphi]$	non-dimensional value of φ	see φ_o	29
$\hat{\varphi}$	$[\varphi]$	unit magnitude of φ		93

1.8   Nomenclature : Differential, Integral and other Math Symbols

Symbol	Units	Quantity	Comments	page
$\equiv$	-	Equivalent to..		15
$[...]$	-	Dimensions of..		18
δf	$[f]$	Infintessimal bit of f	Used in deriving differential/integral equations	11
Δf	$[f]$	Measurable bit of f	Used to define measurable changes	11
$\partial/\partial x$	m^{-1}	Partial derivative	Eulerian quantity, w.r.t. time also	11
d/dx	m^{-1}	Total derivative	Eulerian quantity, w.r.t. time also	11
D/Dt	s^{-1}	Material Derivative	Lagrangian quantity, sometimes called substantial, total or Lagrangian derivative	14
$\oint dZ$		Integral of property Z over a cycle		12
$C\vert_X$	n/a	"C at x=X"	Sometimes written C(x=X)	28

1.9   Non-Dimensional Numbers

Understanding what non-dimensional numbers are telling you is the way smart engineers understand thermofluids. You have an easy way to predict what type of physical system is present, without doing any work. Unfortunately, there are a near infinite number and the various engineering disciplines have different ways of representing them, which is a source of much confusion to the uninitiated. However only a handful are in common use. Those used in this book are listed below, with a page reference to show an example of their first use.

Symbol	Quantity	Equation	Comments	page
Eu	Euler	$p/(\rho u^2)$	Ratio of pressure to kinetic energy	20
Fr	Froude Number	$u/(gx)$	Ratio of Gravitational to Convective Force	20
Ma	Mach Number	u/a	Ratio of Pressure wave speed to Flow Speed	116
Pe	Peclet Number	$\rho C_v ux/k$	Ratio of Convective to Diffusive Flux	29
Pr	Prandtl Number	$\mu C_v/k$	Ratio of Diffusive Momentum and Thermal Flux	29

Sc	Schmidt Number	$\mu/(\rho D_{ab})$	Ratio of Diffusive Momentum and Species Flux	116
Re	Reynolds Number	$\rho u x/\mu$	Ratio of Convective to Diffusive Momentum	20
Ca	Cavitation Number	$(p - p_v)/(\rho u^2)$	Local pressure above fluid vapour pressure relative to kinetic energy	61
C_f	Friction Coefficient	$\tau/(\rho u^2/2)$	Ratio of Viscous Drag to Convective Force	79
C_p	Pressure Coefficient	$\Delta p/(\rho u^2/2)$	Ratio of Pressure Drop to Convective Force	87
C_d	Drag Coefficient	$F_D/(\rho u^2 x^2/2)$	Ratio of Total Drag Force to Convective Force	19

Further information first at section 2.4.5 on page 20, then throughout the subsequent chapters to section 8.8 on page 115. As noted there, this section lets you understand what non-dimensional numbers are relevant to your current set of physics.

2 | Tools of the Trade

This section outlines all the maths and introduces all the physical concepts in component form you need to know to get through to the end of chapter 5. From chapter 3 on page 44 we will assume you do and will reference back to parts of this chapter to remind you. Remember, the nomenclature for everything used in this book is defined on page 3. The glossary (the specific meaning of important words have that are underlined in the text) is defined on page 1. Hopefully, you can skim a lot of this and dwell a few places.

2.1 Eulerian Coordinate Systems

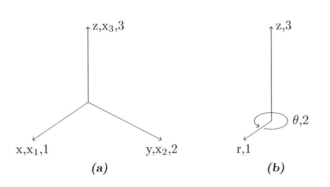

Figure 2.1: (a) Cartesian and (b) Cylindrical Coodinate systems

By Eulerian we mean information (temperature say) at a position relative to some "zero" position or origin. By Coordinate Systems we mean methods to measure space. Coordinate systems are most appropriate when they naturally fit the space, for instance we use a Cartesian (x, y, z) system for a box, a cylindrical (r, θ, z) system for a pipe, as shown in Fig. 2.1. Note there are various ways to represent coordinate directions. As shown in Fig. 2.1a the Cartesian z direction may be defined in terms of z, x_3 or simply the 3 direction. A key point is choice of coordinate system is independent of the physics - if you decided to represent the flow in a pipe in terms of physics defined in the Cartesian coordinate system you have done nothing wrong, just made your job mathematically harder. Indeed a requirement of a correct equation is that you can transform the equation between coordinate systems and the physical process it represents does not change. Coordinate transformations are intimately bound up in the definition of second order tensors, outlined in chapter 6 but you don't have to worry about this until page 88. All you need to know right now is that all the coordinate system axes should be orthogonal.

2.2 Position, Velocity, Acceleration, Force, Energy and Power for a Constant Mass

Here we remind ourselves of the laws of motion in rectilinear and rotational motion, in the Cartesian and cylindrical coordinate systems noted above.

In the Cartesian system, and for the sake of simplicity assuming the x-direction, velocity (u_x) is displacement (x) per unit time (t), momentum (M_x) of an object is its mass × velocity. Acceleration (a_x) is velocity change per unit time, force f_x is mass (m) × acceleration, energy (E) is force × distance, power (P) is energy change per unit time. Force is also the pressure multiplied by the area in the direction of the force. Following on energy = Pressure × Area × distance or pressure × volume, and so on. Objects with a lot of momentum have a lot of inertia need a lot of force applied to get them to stop. Would you rather stop a ping pong ball or a train going at 10 m/s?

$$u_x = \frac{dx}{dt}, \quad M_x = mu_x, \quad a_x = \frac{du_x}{dt}, \quad F_x = ma_x, \quad F_x = pA_x, \quad E = F_x x = pA_x x = pV, \quad P = \frac{dE}{dt} \quad (2.1)$$

We also need to define a few rotational applied maths terms, and to be precise, motion in the θ direction of the cylindrical coordinate system. The <u>rotation rate</u> (or angular speed) $\dot{\theta}$ is the rate of rotation about a point and is given in radians per unit time. The rotational moment is commonly known as the <u>torque</u> (T_θ) and is a product of the force in the direction of rotation F_θ and the distance of the force from the origin r. The rotational force is the <u>angular momentum</u> M_θ and the <u>angular velocity</u> u_θ is best thought of as the angular momentum per unit mass, though many texts refer to it directly as the rotation rate.

$$T_\theta = rM_\theta, \quad M_\theta = mu_\theta, \quad u_\theta = \frac{M_\theta}{m}, \quad \dot{\theta} = \frac{d\theta}{dt} \quad (2.2)$$

2.2.1   Scalars and Vectors

<u>Scalars</u> are an information "packet" that has one "bit" of information –magnitude: familiar examples are pressure, temperature, the number of pints of beer you drank last night, the magnitude of your hangover this morning. <u>Vectors</u> are an information "packet" that has two "bits" of information - magnitude and direction. Each vector has 3 elements of information - one for each coordinate direction. If I said the car is moving at 20 m/s, then I am defining scalar information, speed, but no direction. However, if I said the car is moving at 20 m/s *towards you*, then that is (for you anyway) a different matter entirely. We use vectors to define the direction of coordinate system axes. For instance in Fig. 2.2 $\vec{e_1} = [1, 0, 0]$ defines the direction of the x-axis. It is a displacement vector from the origin at (0,0,0) to a position (1,0,0), in the x-direction only.

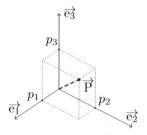

Figure 2.2: *Graphical Representation of point $\vec{P}$*

Generally the position of a point in a Cartesian coordinate system is a vector, $\vec{P} = [x_1, x_2, x_3]$, it gives direction and distance from the origin. Looking at equations 2.1 that some of these equations are scalar equations (e.g. $P = dE/dt$) and some are really vector equations (e.g. $\vec{u} = d\vec{x}/dt$). Therefore, when we see a vector equation, we are really seeing a shorthand way of writing 3 equations, one for each component direction.

Another way to represent a vector is in terms of scalar magnitudes and axis vectors as $\vec{P} = p_1\vec{e_1} + p_2\vec{e_2} + p_3\vec{e_3}$. Note the magnitude of the axis vector $\vec{e}$ is unity, and hence are termed unit vectors. So, back to our car, the speed (a scalar) is 20m/s, the velocity (a vector) might be (-20,0,0). We can also define the magnitude of a vector as $|\vec{P}| = \left(p_1^2 + p_2^2 + p_3^2\right)^{1/2}$. As shown in Fig. 2.2 this is the scalar distance from the origin to point $\vec{P}$, in 2D this is simply Pythagoras's rule. Strictly, as defined below by equation 2.3 in section 2.2.3, the vector magnitude is obtained by taking the dot product of the vector with itself and then taking the square root of this.

2.2.2   Scalar and Vector Fields

In the above section we were implicitly defining scalars and vectors at a single point in space. For instance we could define the temperature at point P, $T(\vec{P})$ or the velocity $\vec{u}(\vec{P})$. The latter reads as 'the vector $\vec{u}$ at point $\vec{P}$'. The real power of scalar and vector information is in terms of fields (of points). Fields can be one, two or three (space) dimensions, for instance the temperature distribution in a steel bar insulated along its length would be a 1-D temperature field $T(x)$ and that change in temperature with position tells us something to do with the rate of heat flow along the bar.

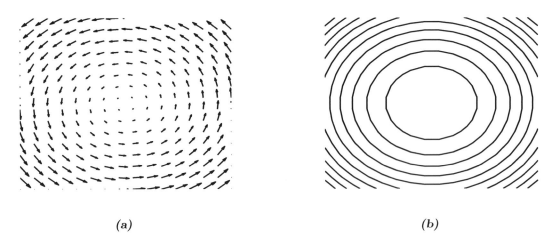

(a)   (b)

Figure 2.3: *a vector (a) and a scalar (b) field*

The key point is that spatial variations in scalar and vector information can give physical insight in how quickly information is being moved through the domain. Fig. 2.3a shows the vector field from a vortex and we can see the rotation is counter-clockwise. When we take the dot product of the velocity vector with itself and take the square root of this, we obtain the speed field, the scalar magnitude of the velocity vector field. This is shown in Fig. 2.3b as a contour plot and we have lost all the information concerning the flow direction, we only now know the magnitude.

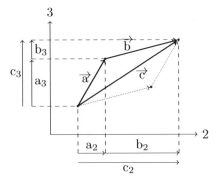

Figure 2.4: *Vector Addition*

2.2.3 Scalar and Vector Operators

An <u>operation</u> is something we do to a variable or pair of variables to change them. For instance *1+1=2* is an addition operation. We are familiar with the standard scalar operators, $+$, $-$, $/$, $\times$. All of these apply to vectors also, with the exception of division (by a vector). Vector operators are easy to visualise, as shown in Fig. 2.4. Addition is defined by $\vec{a} + \vec{b} = \vec{c}$, subtraction similarly and multiplication of a vector by as scalar as $K(\vec{a} + \vec{b}) = K\vec{a} + K\vec{b}$. The other key vector operator you need to be acquainted with is the <u>dot product</u>, which is defined by eqn. 2.3, because $\vec{e}_1 \cdot \vec{e}_1 = 1$ and $\vec{e}_1 \cdot \vec{e}_2 = 0$ (because the axes are orthogonal).

$$\vec{a} \cdot \vec{b} = a_1 b_1 + a_2 b_2 + a_3 b_3 = |\vec{a}| \left|\vec{b}\right| \cos \theta \tag{2.3}$$

The dot product produces a scalar from two vectors and, with reference to Fig. 2.5 in words it is "how much of a is in the direction of b". Looking at Fig. 2.5 when $\theta < \pi/2$ then some of $\vec{a}$ is in the direction of $\vec{b}$ and the magnitude is positive. When $\theta = \pi/2$ none of it is, and when $\theta > \pi/2$ then $\vec{a}$ is going in the wrong direction and the magnitude is negative. It is used in three instances. The first is defined above, defining the vector magnitude, and this can be seen from some of the basic equations (equations 2.1). For instance the scalar energy is really the dot product of the vector force and vector displacement, $E = \vec{F} \cdot \vec{x}$. The second is to define force components in a given direction, where vector $\vec{a}$ (say) is the direction you want and the scalar answer is the magnitude of $\vec{b}$ in the direction of $\vec{a}$. Another use of this is page 69. The third use to define the mass flow through a surface when the flow is not normal to the surface. Then one of the vectors is the unit normal to that surface, see page 63. This is all you need to know until the end of chapter 5. For the keen, this is the start of very powerful mathematical language, tensor notation (see chapter 6) and used extensively to define the

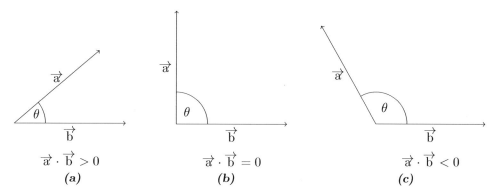

$\overrightarrow{a} \cdot \overrightarrow{b} > 0$ $\overrightarrow{a} \cdot \overrightarrow{b} = 0$ $\overrightarrow{a} \cdot \overrightarrow{b} < 0$

(a)   *(b)*   *(c)*

Figure 2.5: *Graphical Representation of Dot Product Magnitudes*

fluid conservation equations, see chapter 8.

2.2.4 Infintessimal and Finite Increments. Limits. Total Derivative

Let us say we have a curve and we want to investigate the change in y with respect to the change in x. In Fig. 2.6, point 1 is our fixed point. If all we are interested in is the change in y between points (say 1 to 4) then Δy is the symbol used here, though to be specific we usually use Δy_{14} or just y_{14}. *These are finite measurable changes, i.e. things we could see.*

Sometimes we need to derive equations based on very small changes, here we would represent a change in y by δy. Looking at the graph, as point 4 moves to 3, 2 and on to 1 the change in δy becomes smaller and smaller. In the limit of $\delta y \to 0$ and $\delta x \to 0$ at point 1 (say) then $\delta y/\delta x \to dy/dx$. This is the gradient of y with respect to x AT A POINT. More formally, $dy/dx = \lim_{\delta y \to 0} \delta y/\delta x = (y(x+\delta x) - y(x))/\delta x$. This is known as a <u>total derivative</u>. The change in y is *totally* defined by the change in x. Usually we generate a derivative with respect to a coordinate axis - sometimes we might derive a change along a curve. Derivatives are incredibly important in thermofluids - they can be considered driving forces for <u>transport</u> processes. The first order derivative is commonly known as a <u>gradient</u>, it tells use how much a variable changes per unit of space or time. In section 2.6.2 on page 23 we show that heat flows due to a temperature gradient. Mass usually flows due to a pressure gradient, as outlined on page 60.

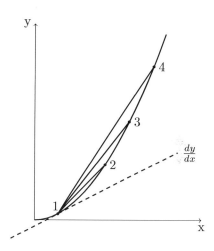

Figure 2.6: *A Geometrical Representation of (total) Derivatives*

2.2.5 Partial Derivatives

In section 2.2.4 we defined the rate of change of y (the dependent variable) in terms of x (the independent one). The example used is heat conduction down a steel bar, and relates the temperature change to the distance along the bar. The temperature in this case is totally defined by the position along the bar and thus T as a function of x is a *total* derivative. Now let us take a more complicated case, and we say that the temperature distribution in the bar is a function of both position along the bar, but also time. Figure 2.7 might represent this unsteady 2D heat conduction process, where y represents t and f represents T. Purely in terms of the maths we have some dependent variable f that is a function of more than one independent variable. In this case let's say two, x and y. For the total differential change in $f(x, y)$ for changes in x **and** y consider a small part of the surface, shown in Fig. 2.7, where f changes by δf and x and y by δx and δy. Since $\delta f = f(x + \delta x, y + \delta y) - f(x, y)$ then we can also write

$$\delta f = \left[\frac{f(x + \delta x, y + \delta y) - f(x, y + \delta y)}{\delta x}\right] \delta x + \left[\frac{f(x, y + \delta y) - f(x, y)}{\delta y}\right] \delta y$$

Taking the limit as before, $\delta x \to 0$ and $\delta y \to 0$

$$df = \left[\frac{\partial f}{\partial x}\right] dx + \left[\frac{\partial f}{\partial y}\right] dy \tag{2.4}$$

where $\partial f/\partial x$ and $\partial f/\partial y$ are the partial derivatives of f with respect to x and y. The former is at constant y and the latter is at constant x, often written $\partial y/\partial x|_x$. The total change with respect to x is df/dx, which can be obtained directly from equation 2.4. Until the end of chapter 5 partial derivatives are used to define equations and also the specific heats (section 2.7.5 on page 32). They are used extensively in chapter 8 deriving the full conservation equations.

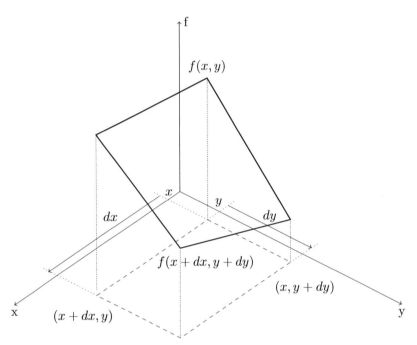

Figure 2.7: *A Geometrical Representation of Partial Derivatives*

2.2.6   Taylor Series

A Taylor series is used to take known information, temperature say, at one point (in time or space) and use this to estimate this information at a nearby point in time or space. The known information includes first and higher derivatives. As shown in Fig. 2.6, let us say we know what y, dy/dx, d^2y/dx^2.... is at $x = x_1$. We use this information to work out what $y(x_2)$ is for instance. A Taylor series is defined : $y(x_2) = y(x_1) + (x_2 - x_1)\frac{dy}{dx}\Big|_{x_1} + \frac{(x_2 - x_1)^2}{2!}\frac{d^2y}{dx^2}\Big|_{x_1} \cdots \frac{(x_2 - x_1)^n}{n!}\frac{d^ny}{dx^n}\Big|_{x_1}$. In theory accuracy improves with the order used, however this assumes we have accurate estimates of higher order terms, which is almost never true. We normally use Taylor series to derive equations, and in this case we take a *very small* interval, and because of this we can use only the first order term, which simplifies our maths considerably.

$$y(x_2) = y(x_1) + (x_2 - x_1)\frac{dy}{dx}\Big|_{x_1} . \tag{2.5}$$

You can see examples of this in section 2.6.5 on page 27 for instance.

2.2.7   Basic Integration Terminology

Very often as engineers we develop a basic equation in terms of an infintessimal change, and then integrate it to obtain a macroscopic change, something we can measure. If we consider a simple piston-cylinder arrangement as shown in Fig. 2.8, we do a small amount of work compressing the gas in the cylinder, $\delta W = pA\delta x = p\delta V$. This defines an infinitesimal change in work (energy) is required to reduce the volume of a gas at a given pressure. The integral effect of these infinitesimal changes results in $W = \int \delta W = \int pA\, \delta x$, the work we get out of an internal combustion engine stroke for instance. As written above, this is an indefinite integral – if you integrate it a constant arises which has to be defined from a boundary condition. The alternative is to use definite integrals, e.g. $W = \int_{V_1}^{V_2} p\,\delta V$. This is however useless for deriving equations and so the former method is preferred.

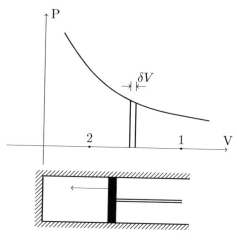

Figure 2.8: *Integrating the pressure-volume change in a piston-cylinder to calculate the work down*

Taking the piston analogy one stage further, considering the motion of a piston in a engine cycle, which starts at one point $x = X$, moves away and then returns to that point. Plotting how pressure and temperature change in the cylinder, for example a two process cycle as shown in Fig. 2.9, ensures that the cycle is closed – it ends where it starts. This has a special notation $W_{net} = A\int_1^2 p\,\delta x + A\int_2^1 p\,\delta x = A\oint p\,\delta x$. As we can see because the two integrals are different the value of this cycle integral is not zero, and is the net work over a cycle. More on this in section 3.11 on page 51 where we discuss engine efficiency. Although you may well be asked to define 2D integrals, you will always be able to simplify them into 1D integrals, for instance to estimate volume flow rate in a pipe, $\dot{Q} = \int u\delta A = 2\pi \int ur\delta r$ as in section 5.4 on page 77. 3D integrals are used for definitions only, for instance the mass of a fluid is defined as the density distribution integrated over a volume. : $m = \int \rho\,\delta V$.

2.2.8 Centre of Area

The centre of area is the the point at which a cardboard cut out of a shape could be perfectly balanced on the tip of a pencil. We only need it once in section 4.9.3 on page 69 for a semicircle so we shall use this as the example. The general equation for the centre of area is $\bar{x} = (\int x\delta A)/A$. Imagine the semicircle is horizontal and the edge is defined by $y = (1-x^2)^{1/2}$, so the origin of the semi-circle is at $(0,0)$ and the area is $A = \pi R^2/2 = \pi/2$. The x-coordinate of the centre of area must be zero. The y coordinate of the *semi*circle is $y_a = (1/A)\int x(1-x^2)^{1/2}dx = 4R/3\pi$. Note this calculation is best done in the cylindrical system with $\partial A = r\partial\theta\partial r$.

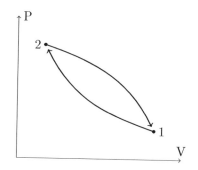

Figure 2.9: *A simple two-cycle process*

2.2.9 Gauss's Law

Gauss's Law is a general law applying to any closed surface (2D) or volume (3D) and in thermofluids it is used to define conservation of information flowing into and out of a volume through its surface. In words it says the rate of change of stuff in a volume is equal to the net (in-out) rate of stuff flowing through the surface of the volume. The classic case is mass conservation, and defines that (as long as no reactions are occurring and the flow is steady) what flows in must flow out. It is used in section

4.6 on page 63.

2.2.10   The Material Derivative: Relating the Eulerian and Lagrangian viewpoints

Imagine it is an autumn day, and you are next to the river - and there are dead leaves floating on the water surface. There are two main ways of representing the motion. One is a fixed frame of reference - you could be sitting on the riverbank [with your neck in brace] watching the leaves float by. This is known as an Eulerian frame of reference (or coordinate system). Everything we have done so far has used this assumption : in the example, the origin of the coordinate system is some spot between your eyes, on the river bank and the (x, y, z) position of the leaf at time t is defined with respect to that. Two positions, at two different times give an Eulerian velocity, two velocities at two different times give an acceleration etc. The other takes the origin as the moving leaf - imagine you are sitting on it, and the only changes you see are with respect to time, for instance if the leaf is speeding up. This is known as the Lagrangian point of view (or frame of reference, or coordinate system).

The basic equations of thermofluids are defined in terms of the Lagrangian point of view, because the rate of change of mass, momentum and energy is defined in terms of a fixed mass. The classic example is Newton's 2nd Law (his apple say), note however just because the Lagrangian viewpoint uses a fixed mass, it does not mean the volume of this mass is fixed in magnitude or shape. This depends on what type of material it is (section 2.5.2 on page 21), and for an apple say, Newton's Law is easy to visualise, however when it is a fixed mass of gas say, of no defined shape or volume, things become a lot more difficult. This is why we as fluids engineers prefer conservation laws based on fixed volumes. Therefore we need a way to relate the Lagrangian and the Eulerian viewpoints. The second complication is that this needs to be done for a fixed mass (and therefore a non-zero volume) because our fundamental equations are written for fixed masses. This requires something called the Reynolds Transport Theorem, and is dealt with in section 2.3.6 on page 17, after a few more definitions are introduced.

First let us relate Lagrangian and the Eulerian viewpoints for a point. If we take a Lagrangian scalar variable φ (the temperature of a small mass moving in a fluid say), then from the Eulerian viewpoint φ is a function of position and time, i.e. : $\varphi \equiv \varphi(x, y, z, t)$. If we now watch that small mass, in a short time period δt the position changes by δx, δy, δz and the value of φ changes by $\delta \varphi$, which may be defined:

$$\delta\varphi = \frac{\partial\varphi}{\partial t}\delta t + \frac{\partial\varphi}{\partial x}\delta x + \frac{\partial\varphi}{\partial y}\delta y + \frac{\partial\varphi}{\partial z}\delta z ...or... \frac{\delta\varphi}{\delta t} = \frac{\partial\varphi}{\partial t} + \frac{\partial\varphi}{\partial x}\frac{\delta x}{\delta t} + \frac{\partial\varphi}{\partial y}\frac{\delta y}{\delta t} + \frac{\partial\varphi}{\partial z}\frac{\delta z}{\delta t}$$

The LHS is the total change in φ in time δt as observed in the Lagrangian frame. In the limit of $\delta t \to 0$ $\delta\varphi/\delta t \to D\varphi/Dt$: this is known as the Material derivative. It represents the total change in φ seen by an observer following the fluid. On the RHS, in the limit of $\delta t \to 0$ $\delta x/\delta t \to u$ and so on for v and w and represents the total change as seen by an Eulerian observer (in the previous example you are on the river bank and the velocity components are that of the leaf). Therefore we can relate the Lagrangian (LHS) and Eulerian viewpoints (RHS) :

$$\frac{D\varphi}{Dt} = \frac{\partial\varphi}{\partial t} + u\frac{\partial\varphi}{\partial x} + v\frac{\partial\varphi}{\partial y} + w\frac{\partial\varphi}{\partial z} \tag{2.6}$$

If we were mathematicians, this is all we would need, because we have related a point in the Lagrangian viewpoint to a point in the Eulerian viewpoint. *but*, a point has zero volume, hence zero mass, and hence Newton's 2nd Law cannot be transformed to an Eulerian viewpoint using this method. As engineers we have two ways to get around this. In section 2.3.4 on page 16 we introduce a system, a flexible volume containing the same fixed mass and on page 17 a control volume, a fixed volume through which a fluid can flow to get around this inconvenience. These in turn are directly related to <u>closed</u> and <u>open</u> processes. <u>Closed</u> processes operate on a fixed mass of fluid, while <u>open</u> processes operate on a fixed volume, through which mass may flow.

2.3 Conservation Principles

Here the tools we use to conserve mass, momentum and energy are introduced. The fundamental basis for conservation of these quantities is based on a fixed mass, however as engineering we generally want to be able to decribe mass, momentum and energy balances in a fixed volume, thus we need to define a conservative way to transfer information between these two frameworks.

2.3.1 Fundamental Conservation Laws

With one very important exception everything is conserved. Our one exception is entropy: as we will discover normally it increases and *is not conserved*. Here we are concerned with three very important conservation laws which you will be use religiously for the remainder of your degree. These are stated in the Lagrangian viewpoint, for a fixed mass, but as we will discover, we prefer them in an Eulerian point of view. The first law in the Lagrangian viewpoint is trivial, the rate of conservation of Mass: $m =$ constant, the rate of Change of Momentum (Newton's 2nd Law), and finally the rate of change of energy. This states that the rate of change of energy of a fixed mass is equal to the thermal energy transferred to it minus the work it does on the surroundings. These are :

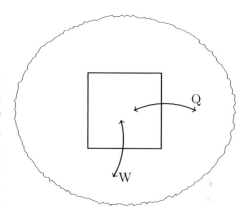

Figure 2.10: *A System exchanging work and thermal energy with an Environment*

$$\frac{Dm}{Dt} = 0 \tag{2.7}$$

$$m\vec{a} = m\frac{D\vec{u}}{Dt} = \vec{F} \tag{2.8}$$

$$\frac{DE_T}{Dt} = \dot{Q} - \dot{W} \tag{2.9}$$

2.3.2 Forms of Energy and Energy Conversion

Energy cannot be created or destroyed, it can only change form. Whilst this is true, energy 'concentration' is also important, as we will discover 1kg of water at 1000K is much more useful than 3kg of water at 333K. A key function of an engineer is to understand and calculate energy budgets, engineers are in the business in making useful energy out of non-useful energy. The forms of energy we have available are:

- Potential Energy : $E_g = mgx$, for example transformed into useful energy by using a dam.
- Kinetic Energy : $E_k = 1/2mu^2$, used to provide thrust to jet engines by a nozzle.
- Thermal (Internal) Energy : $E_u = mC_vT$, used to raise steam to drive turbines in power stations.

There are others but we do not consider them here.

2.3.3 Extensive and Intensive Quantities

The conservation laws of section 2.3.1 are based on <u>extensive</u> quantities, quantities that depend on the amount of mass present.

Mass ($Z \equiv m$)	density ($z \equiv 1$)
Momentum ($Z \equiv m\vec{u}$)	Velocity ($z \equiv \vec{u}$)
Internal Energy ($Z \equiv E_u$)	Specific enthalpy ($z \equiv e_u$)

Table 2.1: *Extensive quantities (L) and their Intensive Equivalents (R)*

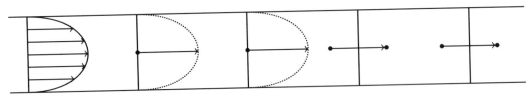

Figure 2.11: *A System (L, dotted lines) and a control volume (R, solid lines) in a pipe*

Our conserved properties are generally <u>intensive</u> properties, independent of the actual amount of A considered. Therefore density (mass per unit volume), velocity (momentum per unit mass), specific energy (energy per unit mass) and species mass fraction (mass of species per mass of continuum) are the norm. Key quantities you need to know are given in table 2.1, and please note the upper/lower case nomenclature.

For any <u>extensive</u> property Z, its corresponding <u>intensive</u> property z may be defined,

$$Z = \int \rho z \, \partial V \approx z \, \Delta V. \tag{2.10}$$

2.3.4   Systems and Environments

A <u>system</u> is a continuous mass of fluid that always contains the same mass (i.e. the same molecules), but whose shape/volume may change. Water sloshing around in a glass is a system. Figure 2.10 gives an example of a system as a box, note energy (work) can transfer across system boundaries but mass (and therefore momentum) cannot. They are by definition a Lagrangian description of a fixed mass. Systems are fundamental to the conservation laws because we conserve information over the same molecules, energy may cross a system boundary (to heat a gas molecule up for instance) but no mass can cross. Equations 2.7 to 2.9 on page 15 are <u>system</u> conservation laws.

Energy transfers take place between a system and the <u>environment</u> it is in. An example is a gas being heated in a piston-cylinder arrangement: heat energy transfers into the system, heats it up, and the system then does work (energy = force × distance) on the environment. Because of the zero flow restriction over the system boundary – thermofluid operations that occur in systems are called <u>closed processes</u> and these are considered in chapter 3 of this book. Because engines work on cycles, they can be approximated by a system analysis (the intake/exhaust processes can be approximated as heat transfers) – in-

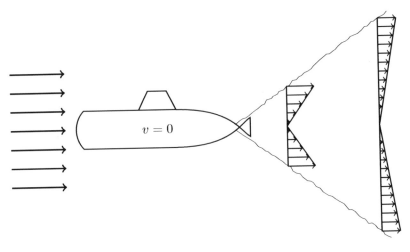

Figure 2.12: *The Wake Structure around a Submarine after a Change of Reference Frame*

ternal combustion engines and gas turbines are considered in section 3.9 on page 49.

2.3.5 Control Volumes

Control volumes do not deform with the flow and mass, momentum and energy are allowed to cross the control volume boundary. They are by definition a fixed volume and thus an Eulerian description. Therefore, examining pipe flow as shown in Fig. 2.11, a system will deform with time whereas a control volume is fixed and flow moves through it. Control surfaces are important for defining mass and momentum transfers across them. One can draw a control volume around a jet engine to calculate the thrust generated for example. Mass/momentum/energy flow per unit area is termed a mass/momentum/energy flux. Because of the relaxation of the flow restriction in control volumes – thermofluid operations that occur in systems are called open processes and these are considered in chapters 4 and 5. The control volume basis is also used directly in computational modelling of thermofluid operations introduced in chapter 14.

2.3.5.1 Moving Control Volumes and Changing the Frame of Reference

The vast majority of the time our control volume is fixed in space, but there are occasions when we want to define a moving control volume - we define a moving frame of reference. In Fig 2.12 we want to examine the flow around a moving object, a submarine. In reality the flow is stationary and the object is moving at a velocity u. As long as the object is not accelerating we can change the frame of reference, make the object stationary and then flow passes over it. In the example we change the frame of reference, i.e. $v = -u$, then our submarine is stationary and the ocean flows past it. The wake is now stationary. Remember to convert the answer back into the correct frame of reference when you have done the question. An example of its use is in section 4.10 on page 70.

2.3.6 Relating Systems to Control Volumes : Reynolds Transport Theorem

As noted in section 2.2.10 on page 14 for a point we have already described how to transfer information from a Lagrangian to an Eulerian coordinate system via the material derivative. This is of no use to develop Eulerian versions of the fundamental conservation equations defined in section 2.3.1 on page 15 because the amount of mass at a point is zero. Conceptually, applying Newton's 2nd Law for a fixed shape and fixed mass (a solid metal sphere say) then this is not a problem. For a liquid our fixed mass does not have defined shape, and a gas *in addition* does not a defined volume. This is why we need to define conservation laws for fixed volumes. The Reynolds Transport Theorem is derived by considering the rate of change of any extensive variable of a system as it passes through a fixed control volume. We will use mass as our extensive variable (density is the intensive equivalent), but we can use any extensive variable, for instance momentum or energy (see table 2.1 on page 16).

In Fig. 2.13 the control volume is defined by the dashed line and the system as the area defined by the dotted area. Also observe that the volume of the system is slightly larger than the volume of the control volume, at time 0 a little bit of the system *mass* has not yet flowed into the control *volume*. Considering first purely from a system point of view : The mass of the system at time 0 is: $m_{sys}(0) = m_{CV}(0) + \delta m_{in}$. And at time δt: $m_{sys}(\delta t) = m_{CV}(\delta t) + \delta m_{out}$. The change in mass of the system is defined by the material derivative: $Dm_{sys}/Dt = \lim_{\delta t \to 0}((m_{sys}(\delta t) - m_{sys}(0))/\delta t)$. Substituting in the above definitions: $Dm_{sys}/Dt = \lim_{\delta t \to 0}(1/\delta t)m_{CV}(\delta t) + \delta m_{out} - m_{CV}(0) - \delta m_{in}$. Re-arranged: $Dm_{sys}/Dt = \lim_{\delta t \to 0}(1/\delta t)m_{CV}(\delta t) - m_{CV}(0) + \lim_{\delta t \to 0}(1/\delta t)\delta m_{out} - \lim_{\delta t \to 0}(1/\delta t)\delta m_{in}$. The following should be noted: 1st term of the RHS is the rate of change of mass *inside the CV*, and is thus an *Eulerian* quantity: $\partial m/\partial t$. Converting this to an intensive quantity: $\partial m/\partial t = (\partial/\partial t)\int \rho\, \partial V$. 2nd and 3rd terms are the net rate of mass flow rate into the CV. If the flow was uniform, then the mass flow rate out of the CV would be $\rho u_n A$, where the u_n is the speed of the flow normal to the CV face. Generalising, the net mass flow into the CV, using Gauss's Law: $\int \rho\,(\vec{u} \cdot \vec{n})\,\partial A$ which gives us

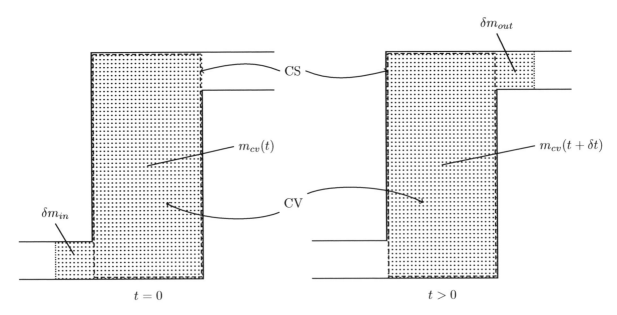

Figure 2.13: *Overlapping Control Volume and System Descriptions*

the final form of the Reynolds Transport Theorem:

$$\frac{Dm_{sys}}{Dt} = \frac{\partial}{\partial t} \int \rho \partial V + \int \rho \left(\vec{u} \cdot \vec{n} \right) \partial A. \tag{2.11}$$

In words, for an extensive (Z), intensive (z) variable pair :

$$\begin{bmatrix} net\ rate\ of\ change \\ of\ Z\ in\ system \end{bmatrix} = \begin{bmatrix} net\ rate\ of\ change \\ of\ z\ in\ CV \end{bmatrix} + \begin{bmatrix} net\ inflow\ of\ z \\ through\ the\ CV\ surface \end{bmatrix}$$

Remember, (Z, z) can be any variable pair, any fluid information. If $Z \equiv$ mass, then $z = \rho$, via equation 2.10 on page 16. Since $m_{sys} = 0$ by definition, $Dm_{sys}/Dt = 0$ which is the fundamental mass conservation law, equation 2.7 on page 15.

2.4   Dimensions, the Buckingham Pi theorem and Non-Dimensional Numbers

2.4.1   Fundamental Dimensions

Fundamental dimensions are the basic building blocks of all variable descriptions and are denoted by [...]. There are only three in use here, Mass [M], Length [L], Time [T]. In general the dimensions of any variable may be defined in terms of a combination of fundamental dimensions $[M]^a [L]^b [T]^c$. Given that all variable dimensions need to be composed of our fundamental building block dimensions, we can immediately relate the familiar dimensions of other variables back to these through use of the defining equations. For example Momentum is mass × velocity and Velocity $u \equiv [L] [T]^{-1}$. Acceleration is the rate of change of velocity with time or $A \equiv [L] [T]^{-2}$. Force is mass × acceleration or $F \equiv [M] [L] [T]^{-2}$. Stress and/or pressure is force per unit area $P \equiv [M] [L]^{-1} [T]^{-2}$. Energy is force × distance $F \equiv [M] [L]^2 [T]^{-2}$.

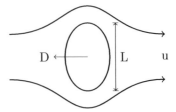

Figure 2.14: *An object in a flow*

2.4.2 Units are not Dimensions

Dimensions of a variable do not change even if the associated units do. For example consider the drag D on an object of diameter L in a flow of speed u shown in Fig. 2.14. The dimensions of velocity are $u \equiv [L][T]^{-1}$. However because there is no scale associated with this image, we could be seeing a microbe or a galaxy. The units used to measure velocity could be light years per hour, mm/s etc. Dimensions define the variable, units define the amount present.

2.4.3 Dimensional Homogeneity

A short but very important section. Checking dimensions during the development of equations and and units when calculating numerical answers respectively is second nature for a competent engineer. Note :

- Every term in any equation, e.g. a, b, c in $a + b = c$ must have the same dimensions. If the equation is dimensionally wrong then it is has no physical meaning, and *must be incorrect.*
- Every term in every equation must also have consistent units. Both in terms of SI versus imperial, but also correct scale. For instance do not use kJ and J or m and ft in the same equation.
- This is an *excellent* way to check your equation derivations and your exam solutions !

2.4.4 The Buckingham Pi theorem

Dimensional analysis allows engineers to combine variables together in non-dimensional groups that are scale independent. Taking Fig. 2.14 as an example, we might wish to define how the drag relates to the speed, object diameter and the fluid properties for example. The key point is we develop a non-dimensional relationship, and this means that we have a way to relate laboratory scale experiments to real world scale. Obviously, if we are concerned about the drag generated by an oil tanker, so we can work out how big an engine to put in it, this is quite useful if we can build a properly scaled model. It is also useful if we have several different experiments of the same process, but with different fluids, oil and air say. Good engineers *always* plot their data non-dimensionally. By plotting all the data from many different experiments on the same graph with non-dimensional axes, they draw a curve and define the *functional relationship* that holds true for *all* the data.

Let us take the example of Fig. 2.14 and say we wish to measure the drag force D exerted on an object of size L in a flow of a given fluid of velocity u. What could/should the drag D depend on? Let us say the Drag $D \equiv [N] \equiv [M][L][T]^{-2}$ depends on: Length $L \equiv [L]$, Velocity $u \equiv [L][T]^{-1}$, Density $\rho \equiv [M][L]^{-3}$, viscosity $\mu \equiv [M][L]^{-1}[T]^{-1}$. In words, the drag is a *function* of these variables, therefore $D = f(L, u, \rho, \mu)$. Let us try to make the LHS of this non-dimensional by dividing the LHS by using some of our variables on the RHS, to make the term have no dimensions. $\rho L^2 u^2 \equiv [M][L]^{-3}[L]^2[L]^2[T]^{-2} = [M][L][T]^{-2}$. So now we can re-write our *dimensional* functional relationship, as another *non-dimensional* functional relationship $D/(\rho L^2 u^2) = g(L, u, \rho, \mu)$. Now we said originally that drag depends on 4 variables, and we have used 3. Therefore the term on the right hand side must (a) contain the unused variable (viscosity) and (b) be non-dimensional (otherwise the equation is not physically correct). So we have to use some or all of the other 3 variables again to make another non-dimensional quantity, $\rho u L$ has the same dimensions as μ, so $D/(\rho L^2 u^2) = h(\rho u L/\mu)$. It turns out the RHS is a very important non-dimensional number, the Reynolds Number, which is discussed in context in section 5.1 on page 75. It defines the flow pattern around the object, and that in turn defines the amount of drag generated. So, if we wanted to do work out what the drag was on an object of $D = 100m$ diameter, we could do this on a $D = 1cm$ test object as long as the Reynolds Number on the real and the test object was the same.

It is important to remember what dimensional analysis can do and what it cannot. It *can* tell you what the non-dimensional groups form the functional relationship. It does *not* give you any information on how these groups inter-depend. A formal procedure for the above is known as the Buckingham Pi theorem, which is :

- Number of Groups = Number of Variables − Number of Dimensions : In the above example the number of variables (excluding D, the variable on the LHS to be made dimensionless) was 4, three dimensions, leaving 1 non-dimensional group to define on the RHS.
- Choose at least 3 variables which together contain all the dimensions present. This is the *repeating group*. In the above example the left hand side is $D\rho^a L^b u^c$ which gives $a = -1, b = -2, c = -2$ and therefore $D\rho^{-1}L^{-2}u^{-2}$.
- Take each of the *remaining variables* and form a non-dimensional group with some or all of the repeating group. In the above example our remaining variable was μ, and our repeating variables are $\rho L U$, so we find the coefficients of $\mu\rho^a L^b U^c$, leading to $a = b = c = -1$ and therefore $\mu\rho^{-1}L^{-1}u^{-1}$.

Non-dimensional analysis is used extensively in engineering, examples in this book include section 2.6 on page 23. An example of plotting data non-dimensionally Figure 5.6 on page 79. The non-dimensional numbers used in this book are also listed in table 1.9 on page 6.

2.4.5 Dimensionless Numbers as Physical Ratios

Here the aim is to emphasise the physical meaning of non-dimensional numbers. Initially, consider the dimensionless numbers that are force ratios. We can define several forces, for instance.

- Inertial or convective Force : $F_u = ma = \rho L^2 u^2$.

- Viscous (shear) force : $F_\tau = \mu(du/dy)A = \mu(u/L)L^2$.

- Gravity force : $F_g = mg = \rho L^3 g$.

- Pressure force : $F_p = pA = pL^2$.

and we can combine them to obtain different dimensionless numbers, for example.

- Reynolds Number : $Re = \dfrac{\text{inertial force}}{\text{viscous force}} = \dfrac{F_u}{F_\tau} = \dfrac{\rho L^2 u^2}{\mu(u/L)L^2} = \dfrac{\rho Lu}{\mu}$.

- Froude Number : $Fr = \dfrac{\text{inertial force}}{\text{gravity force}} = \dfrac{F_u}{F_g} = \dfrac{\rho L^2 u^2}{\rho L^3 g} = \dfrac{u^2}{Lg}$.

- Euler Number : $Eu = \dfrac{\text{pressure force}}{\text{inertial force}} = \dfrac{F_p}{F_u} = \dfrac{pL^2}{\rho L^2 u^2} = \dfrac{p}{\rho u^2}$.

So, for instance, when the Froude Number is really big, we know that the gravitational forces are negligible, and so on. One word of warning : The non-dimensional numbers are sometimes written in different form, for instance the Froude Number is more usually presented as the square root of what is shown above. Likewise, different engineering disciplines can also use different forms.

We can consider the relative strength of diffusion coefficients, for instance all in terms of $[L]^2[T]^{-1}$ the momentum, heat and species terms are $\nu = \mu/\rho, k/(\rho C_p), D_{ab}$ which give the following.

- Prandtl Number : $Pr = \dfrac{\text{momentum diffusion coefficient}}{\text{thermal diffusion coefficient}} = \dfrac{\mu/\rho}{k/(\rho C_p)} = \dfrac{\mu C_p}{k}$.

- Schmidt Number : $Sc = \dfrac{\text{momentum diffusion coefficient}}{\text{species diffusion coefficient}} = \dfrac{\mu/\rho}{D_{ab}} = \dfrac{\mu}{\rho D_{ab}}$.

These numbers tell you how quickly heat and chemical species will diffuse in a fluid relative to the rate of diffusion of momentum.

2.5   Definition of a Pure Single State Fluid (Macroscopic Description)

2.5.1   Continuum Assumption

As noted in section 2.7 on page 29 when considering fluids, we do not define them in a microscopic sense, in terms of molecules, but in terms of *average* molecular measures. For instance if a molecule weighs a mass $M \ kg$, and if there were N molecules in a box of volume $V \ m^3$ then the density of the gas would be $\rho = NM/V$. In defining a density (or any other fluid property, like pressure, temperature, viscosity, thermal conductivity) we have made a <u>continuum approximation</u> and we have assumed our box is the "right size".

By reference to Fig. 2.15, if we choose our box to be too small, we might not count enough molecules to get a good average. In other words, if we measured the density several times we might get several different answers, because the number of molecules in the box varied. If our box is too big instead of capturing a macroscopic change in a fluid property we will average it out. For example the density on one side of the container might be different to the density on the other side of the container, if for instance a temperature gradient existed. Solids, liquids and gases have densities commensurate with the spacing between their molecules. For instance steel, water and air have densities of 7700,1000 and 1 kgm^{-3}.

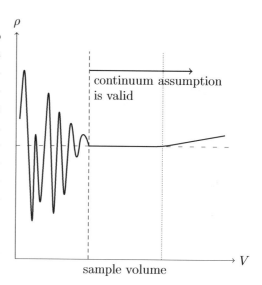

Figure 2.15: *Measuring Density with different sample volumes*

2.5.2   Definition of a Fluid

Solids have molecules that do not move relative to one another, and the molecules are very close together and have strong attractive forces between them. Generally when you apply a force or a stress (a force per unit area) the solid does not deform or if it does, it stops deforming when you remove the force. Liquids have molecules close enough together for attractive forces to dominate and keep a given mass occupying a given volume, but has no preferred shape. Gases have molecules far apart and they have no attractive forces between them and expand to fill any available volume, and has no preferred shape. A fluid is defined as a material that undergoes continuous deformation (straining) due to a shear force (or stress) applied to it. In the fluids we will discuss, the stress and the strain are *proportional* and the fluid is said to be <u>Newtonian</u>.

2.5.3   Response of Solids and Fluids to a Normal Stress

If we impose a normal stress – as shown in Fig. 2.16 a weight on a piston acting in the direction of gravity in two cylinders, one containing a fluid and one containing an elastic solid. In both cases the molecules of the material next to the wall stay in contact with their respective molecules of the wall material. For the solid, let us assume the solid and the wall surfaces are glued together. When the weight is applied the same thing happens: Both the solid and the fluid do the same thing: the solid and the fluid compress a little bit (but by different amounts) and then reach a new equilibrium position. The conclusion is that solids and fluids behave in a similar way when subjected to a normal force (more specifically a normal *stress*, force per unit area). A compressive force in the fluid and the solid balances the weight on the piston and this is the static equilibrium state, where nothing is moving.

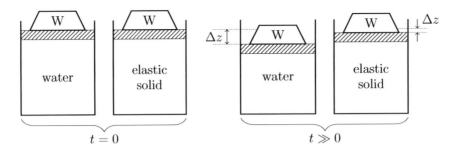

Figure 2.16: *strain response of a solid and a fluid to a normal stress*

2.5.4 Response of Solids and Fluids to a Shear Stress

Now consider two flat large plates parallel to each other separated by a narrow gap. In one case the gap is separated by a viscous fluid, treacle say. The other is our elastic solid. We now apply a shear force – by moving the plates in different directions normal to their orientation, as shown in Figure 2.19 on page 24. The solid behaves as before, it gives a little and then acquires a new static equilibrium, with the stress in the deformed solid balancing the shear force created by the plate motion. The fluid however allows the plate to move continuously as long as the shear force is applied and the equilibrium state (here when the stress in the fluid no longer changes with time) is where the plate and the fluid are in continuous motion. Note however that the relative motion of the wall and the fluid *at* the wall is zero. The motion stops only when the motion of the wall, due to the force applied to the plate, ends. The conclusion is that solids and fluids behave differently under shear. Also shear stresses in the fluid are *only* present when the fluid is in motion. In solids shear stresses remain even when the solid in not deforming and continue to resist the shear stress on the plate. Solids under shear stress behave like solids under normal stress.

2.5.5 Material Compressibility: Bulk Modulus

In reality all materials are compressible to some extent, because all materials at the end of the day are collections of molecules that respond to forces placed upon them. The bulk modulus, K, measures the change in volume arising from a compressive stress. In words it is "the specific change in volume for a given stress". If we say δp is a small change in normal stress (acting in the same direction as the area normal) then this produces $\delta v/v$, a small change in a specific volume. The bulk modulus is therefore defined, $K = -\delta p/(\delta v/v) = \delta p/(\delta \rho/\rho)$. This is because $m = \rho v$, $\delta m = \rho \, \delta v + v \, \delta \rho = 0$ and therefore $-\delta \rho/\rho = \delta v/v$. Liquids have very large K, and hence are virtually incompressible. Typical bulk moduli for steel, water and air are $\sim 10^{11}, 10^9$ and $10^5 Nm^{-2}$ respectively. However if the density changes in the gas are small, then gases can also be treated as incompressible. Note from the gas law (eqn. 2.17 on page 32), for a constant temperature compression, $dp/d\rho = RT$ and $K = (\rho \delta p/\delta \rho) = \rho RT = p$. So the bulk modulus is the pressure for a constant temperature gas.

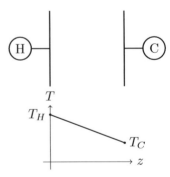

Figure 2.17: *Diffusional Heat Flow due to a Temperature Gradient*

2.5.6 Thermofluid Variables

We are interested in how fluids exert force, move and change energy from one form to another. We therefore have groups of variables that you may have encountered previously in applied maths and physics. These are, pretty much, all the variables we use in this subject.

2.5.6.1 State Variables
These are variables that define the state of a fluid, not the flow, or how it conveys of converts mass, momentum or energy. For a compressible gas density, pressure, entropy, temperature. ρ, p, s, T completely define the gas state. For incompressible liquids, the pressure is irrelevant, and density and temperature are independant.

2.5.6.2 Microscopic Transport Variables
Variables associated with the way in which mass and energy *diffuse* through the stationary fluid due to molecular interactions. These include the viscosity, thermal conductivity, species diffusivity, heat capacity, gas constant, μ or $\nu = \mu/\rho$, k, D_{ab}, C_p, C_v, R.

2.5.6.3 Macroscopic Transport Variables
Variables associated with force and energy transfer in fluids, and often related with fluid motion such position, velocity, acceleration, force, stress, energy, power. These variables are connected via relationships define by equation 2.1 on page 9.

2.5.7 Other Material Properties
There are of course, many other material properties that can be defined. To the end of chapter 5 we only require the vapour pressure, in section 4.3.1 on page 61. This is pressure at which the liquid turns to vapour (boils) and is a strong function of temperature. Therefore, we can make a liquid boil by either heating it up, or reducing the pressure of the air above it. This is why the Nepalese use pressure cookers to cook rice.

2.6 Transporting Fluid Information in Space and Time

When we use a <u>system</u> basis (section 2.3.4 on page 16), our only rate of change is with respect to time. As engineers, we usually want to define the rate of change of a fluids mass, momentum and energy in space. Or, from another point of view we wish to understand how fluid information is *transported* from one location to another.

2.6.1 Convective Transport
Convective transport is the action of the flow to move fluid information from one place to another. If you turn the hot tap in your house on the boiler starts, heats up the water, and then the flow convects this heat in the fluid to the tap. No flow, no convective transport. It is a macroscopic quantity and only occurs in materials that flow, like fluids.

Imagine we have a pipe of cross-sectional area A and the incompressible fluid flow velocity profile across the pipe radius from centreline to pipe wall is constant. Then the mass flow along the pipe would be $\dot{m} = \rho A u$. Now imagine that this *fluid* is hot, and the *flow* is convecting thermal energy from one end of the pipe to the other, and it is doing that at a rate of mass flow (kgs^{-1}) × specific internal energy (Jkg^{-1}), $\dot{m}e_u = \rho A u e_h$. The flow rate of any convected variable other than mass is the mass flow multiplied by the intensive form of that variable.

2.6.2 Diffusive Transport of Scalars
Diffusive transport is the transport of information due to molecular interactions, it does not need a flow to be present to occur. If you hold one end of an aluminium bar and put the other end in the fire, your hand gets hot due to the diffusive transport of thermal energy along the bar. At a microscopic level, when high speed molecules hit low speed molecules, on average, some kinetic energy is donated to the low speed molecules. So, overall thermal energy *always* moves from hot regions to cold regions.

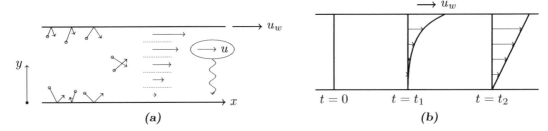

Figure 2.18: *A (L) microscopic and (R) macroscopic interpretation of diffuse momentum transport*

All diffusion processes operate on a <u>gradient</u>, it always a *spreading* process.

Diffusive transport applies to all materials - solids and liquids much better thermal conductors than gases. Using thermal energy as an example, Fig. 2.17 shows a stationary fluid placed between a hot and a cold surface. Molecules pick up energy (speed) from the hot wall and molecules donate energy (speed) to the cold wall. In the macroscopic sense (with a big enough box) the average speed is represented by temperature. Thus there is a gradual change in temperature as, on average, fast molecules from the left collide with slower ones on their right. On average molecular kinetic energy (temperature) *diffuses* from right to left, a thermal energy flow. It turns out that the gradient is linear in Cartesian coordinates. The proportionality between the specific heat flux (W/m^2) is defined by the thermal conductivity of the material. Minus sign because the flux is in the opposite direction to the gradient (i.e. stuff flows downhill). Similarly, the diffusion of a chemical species, m_a, gives

$$\dot{q}_z = -k\frac{dT}{dz} \quad \dot{q}_z = -\rho D_{ab}\frac{dm_a}{dz} \tag{2.12}$$

The k of eqn 2.12 is the thermal conductivity, or the thermal energy diffusion coefficient. Back to our aluminium bar you are holding in the fire, if you exchanged it for a wooden one, the k value drops significantly, and your hand does not get as hot. Likewise, D_{ab} is the diffusion coeffcient of species a in material b. These two diffusion laws are known as Fourier's and Fick's Law, for heat and mass respectively.

2.6.3 Diffusive Transport of Momentum

The transport of momentum (a vector) by diffusion has the same mechanism as scalars such as thermal energy, molecular interactions, but is a conceptually a little more difficult. Obviously the fluid must be in motion, otherwise there is no momentum to diffuse! In Fig. 2.18 the top wall is moving with a speed u_w and the bottom wall is stationary. Molecules are randomly bouncing around as before but on the top wall, every time they collide with it they pick up some momentum (u_w per unit mass) from it.

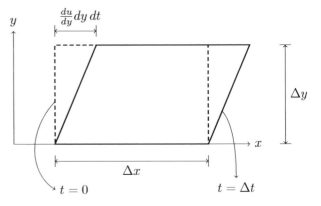

Figure 2.19: *Graphical Representation of Rate of Strain*

This momentum change at the top wall is balanced by a force applied to the wall. These high speed molecules on the top wall get fired into the volume, and when they collide with another molecule they transfer some momentum in the u-direction to it. Over lots of collisions, the molecules at the top have lots of momentum in the u-direction, the ones in the middle have some, and the ones at the

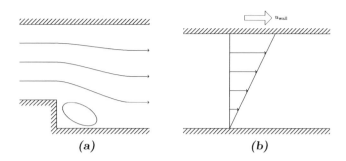

Figure 2.20: *Examples of (a) 2D and (b) 1D flow*

bottom have least of all. If you average the particle speed in horizontal "slices" you find a gradual decay of mean horizontal molecular speed from the top to the bottom wall. So, momentum in the x-direction is added to the fluid at the top wall (in the direction of wall motion) and it "diffuses" into the fluid lower down in the y-direction. The law of momentum diffusion needs to relate the force at the top wall to the macroscopic motion induced in the fluid. The amount of force applied to the top wall may be related to the amount of deformation (strain) in the fluid as shown by Fig. 2.19.

First of all we need to define the velocity gradient, the rate of change of velocity in the x-direction with respect to the change in the y-direction. This is also Rate of Shear Strain of the velocity field and is the deformation in x per unit length in y, per unit time, or $\frac{\frac{dU}{dy}\delta y \delta t}{\delta y}/\delta t = \frac{dU}{dy}$. The force at the top wall is the shear force per unit area, the shear stress in the x-direction. It is *linearly* proportional to the velocity gradient in the y-direction through the viscosity coefficient. Most fluids obey this Newtonian stress-strain relationship, and all fluids considered in this book do.

$$\tau = \mu \frac{du}{dy} \qquad (2.13)$$

μ is known as the "dynamic viscosity", sometimes you might see $\nu = \mu/\rho$ which is known as the kinematic viscosity. Viscosity diffuses momentum like thermal conductivity diffuses heat. As noted previously in section 2.4.5 on page 20 the Prandtl and Schmidt numbers define the ratio of the diffusion coefficients of heat and viscosity and mass diffusion and viscosity respectively.

2.6.4 Fluid Transport Terminology: Boundary Conditions
Before we derive how information about a fluid (i.e. its temperature, density, velocity etc) is transported, we need to define some terminology to define the state of information at the edges of the domain, for instance at the surface of a wall.

2.6.4.1 Steady, Fully Developed and Uniform Flow
If the flow is said to be steady then there is no change from one time point to the next. Therefore $\partial/\partial t\,(\ldots) = 0$. If the flow is said to be fully developed then the fluid velocity and all other variables are constant in the main flow direction. The pressure gradient is also constant (and can be zero). Note: the key point is that the pressure *gradient* is not changing with x. Example: if the flow is fully developed in the x-direction then $\partial/\partial x\,(\ldots) = 0$ except $dp/dx = C$ where C might be 0. If the flow is said to be uniform then the velocity field is constant in space.

2.6.4.2 One or Two Dimensional Flows
If the geometry is said to be large in one or more directions then nothing changes in that direction, meaning that the gradients in that direction are zero. Note: this does not mean the velocity component is zero in that direction necessarily. If the flow is one dimensional and in one of these *large* directions, then the flow will be fully developed. In Fig. 2.20 we see a one and a two dimensional flow, which is large in the z-direction. For the two dimensional flow $u = f\,(x, y)$, $v = f\,(x, y)$ and for the one

dimensional flow $u = f(y)$. In both cases we can safely assume that $\partial/\partial z (\ldots) = 0$, and in the one dimensional case in addition $\partial/\partial x (\ldots) = 0$.

2.6.4.3 Wall Boundary Conditions for Velocity

At a solid surface the no-slip condition applies if the fluid has non-zero viscosity, and this is *always* true in reality. The velocity tangential to the surface, at the surface, is the same as the wall velocity. For the 1-D example in Fig. 2.20 the tangential velocity component is $u = u_{wall}$; $w = 0$ at $y = h$ and $u = 0$; $w = 0$ at $y = 0$. For solid walls, the normal velocity is also zero. For the 1-D example shown in Fig. 2.20 $v = 0$ at $y = 0$ and $y = h$. For porous walls, the normal velocity is defined, in this example, $v = v_{wall}$ at $y = 0$ and $y = h$.

2.6.4.4 Wall Boundary Conditions for Scalars (eg Temperature)

Common scalars one needs to ascribe boundary conditions for are temperature and species mass fraction. These are diffusion type boundary conditions, where the flux through the wall, at the wall is defined by eqn. 2.12 on page 24. Here k is the diffusion coefficient (e.g. thermal conductivity for energy) and is a material property. There are three types of scalar boundary conditions commonly used, which are applied to equation 2.12 on page 24.

- The first type is constant value, $T = const$, for instance a wall held at a fixed temperature. Heat will flow in and out depending on whether the temperature of the fluid is lower or higher than that of the wall.
- The second type is a constant flux, $\dot{q}_x = const$, for instance a constant amount of heat is injected into the domain. The temperature at the wall is not fixed, but determined from the temperature of the fluid near the wall, and the gradient (defined at the wall by the boundary condition).
- A special case of this type is the zero flux condition, $\dot{q}_x = 0$. This assumes (for temperature) a completely insulated wall.

2.6.4.5 Boundary Layer Edge Condition

Near the edge of an external boundary layer it merges into the free stream flow field, and as shown in Fig. 5.10 on page 84 the velocity approaches the free stream velocity $u(y) \to u_0$ as $y \to \infty$. It also means that $du/dy \to 0$ as $y \to \infty$. Note this does not allow to work out where the edge of the boundary layer actually is, for this the assumption $u(y) = 0.99u_0$ is usually made.

2.6.4.6 Symmetry Plane Boundary Condition

A symmetry plane is present if the solution can be reflected across it, like a mirror. As example is shown in Fig. 2.21, the pressure driven flow between two large parallel plates separated by a distance h. There is a symmetry plane at the mid-point parallel to the two plates. The following conditions apply at a symmetry boundary. Zero gradients normal to the boundary, at the boundary, for instance $dT/dy|_{y=h/2} = 0$. Zero normal velocity at (through) the boundary, $v|_{y=h/2} = 0$. Zero tangential strain and stress at the boundary, $du/dy|_{y=h/2} = 0$. It is generally preferred to avoid imposing symmetry conditions for analytical solutions because you are placing a boundary inside the domain. They are

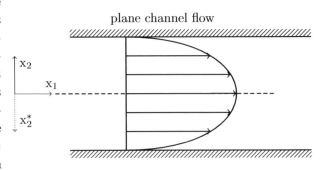

Figure 2.21: *Symmetry Boundary Conditions*

however often used in computational fluid dynamics (chapter 14) to reduce the size of the domain and hence the computer workload.

2.6.4.7 Cylindrical Coordinate System Axis Boundary Condition

Cylindrical systems have an axis at $r = 0$ as sketched in Fig. 2.22. The following boundary conditions apply at the axis ($r = 0$). $u_1 = u_r = 0$, $u_2 = u_\theta = 0$, $\partial u_3/\partial r = \partial u_z/\partial r = 0$.

2.6.5 Balance of Convective and Diffusive Transport in 1-D Uniform flow

Figure 2.22: *Axis Boundary Conditions*

Until now we have discussed how information (such as heat and momentum) is transported in a fluid by convection and diffusion. Understanding the relative strength of convective and diffusive flow is important for a thermofluids engineer because the physical characteristics of the fluid and their numerical implementation in computer models are very different for these two transport mechanisms. This understanding is achieved by non-dimensional analysis and is the subject of our analysis is shown in Fig. 2.23. We have a long channel in the x-direction between two plates that extend very far in the x and the z-direction. We assume that our fluid is inviscid, and hence nothing changes in the y direction either. We are interested in a section of this channel from $x = 0$ to $x = 1m$. The velocity is in the x-direction only $u = u(x)$, $v = w = 0$. The flow is steady and the fluid is incompressible. At $x = 0$, $T = 273K$ and at $x = 1$, $T = 373K$. Imagine we have a pair of perfectly porous meshes held at two temperatures that the fluid can flow through. The transported information we are interested in are mass and thermal energy (temperature). Notice that convection is trying to "blow" 'cold' thermal energy at $273K$ from left to right whilst diffusion is trying to spread 'hot' thermal energy from right to left. This suggests the temperature profile over the range $1 > x > 0$ will be dependent on the relative strength of convection and diffusion.

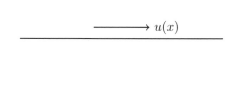

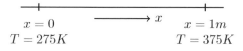

$x = 0$ $x = 1m$
$T = 275K$ $T = 375K$

Figure 2.23: *Setup of the 1-D convection-Diffusion Test Case*

We now consider, as shown by Fig. 2.24 a very small control volume in this channel, at $x = X$. We assume we know everything about the fluid and the flow at $x = X$, e.g. $u(x = X)$, $T(x = X)$, $du/dx(x = X)$, $dT/dx(x = X)$.

Because nothing is changing in the y or z-directions all gradients are zero in these directions. Therefore to conserve any information being transported through the control volume shown all we have to do is work out what is coming in the left hand face and subtract that from what is going out the right hand face. To do this we need to employ a Taylor Series (eqn. 2.5 on page 12) to estimate information at the control volume faces. On the left hand face $T|_{x-\delta x/2} = T|_x - (dT/dx)|_x (\delta x/2)$ and on the right hand, $T|_{x+\delta x/2} = T|_x + (dT/dx)|_x (\delta x/2)$. As discussed in section 2.2.4 on page 11 we are using d/dx here as T depends only on x.

2.6.5.1 Conservation of Mass

For this problem, the flow is steady, and there is no creation or destruction of mass in the control volume, hence the net mass flow of mass across the surface of the control volume is zero. We have assumed we know everything at some point point $x = X$, the control volume centre shown in Fig. 2.24 the mass flow through the area is $\dot{m} = \rho u A = \rho u \, \delta y \, \delta z$. Initially we make no assumption about the compressibility of the fluid. Therefore the mass flow at $x + \delta x/2$ is $\dot{m} = \rho A u|_{x+\delta x/2}$ requires a Taylor expansion for the product ρu. The Taylor series expansion for the upstream face in terms of what we know at $x = X$ is $\rho u|_{x+\delta x/2} = \rho u|_x + (d(\rho u)/dx)|_x (\delta x/2)$. Similarly the Taylor series expansion for the downstream face is $\rho u|_{x-\delta x/2} = \rho u|_x - (d(\rho u)/dx)|_x (\delta x/2)$. Net mass accumulation out − in: $(d(\rho u)/dx)|_X \delta x \delta y \delta z = 0$. If the fluid is incompressible, which we will now assume, then $du/dx = 0$. Notice that for constant density flows (assumed here), velocity cannot change with x.

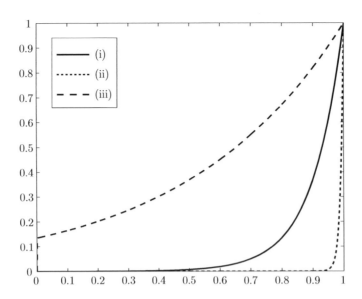

Figure 2.25: *Convective-Diffusive Transport at different Peclet Numbers*

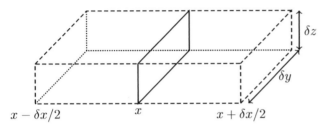

$x - \delta x/2$ $\qquad x \qquad$ $x + \delta x/2$

Figure 2.24: *Control Volume for Flux Discretization*

2.6.5.2 Conservation of Thermal Energy due to Convection

As noted in section 2.6.1 on page 23 the flow rate of any convected variable other than mass is the mass flow multiplied by the intensive form of that variable. For the internal energy flow rate (Js^{-1}) = mass flow (kgs^{-1}) × specific enthalpy (Jkg^{-1}). Given the known internal energy flow at $x = X$: $\dot{m}e_u = \rho A u e_u = \rho A u C_v T = \rho u C_v T\, \delta y\, \delta z$. Since we have assumed our fluid is incompressible, only T varies, u, $C_v = C_p = C$ here, ρ are constant in x. As we will discover (section 2.7.8 on page 33) since the fluid is assumed to be incompressible, the distinction between C_v and C_p is meaningless. Taylor series expansions for the temperature for the upstream and downstream faces are respectively, $T|_{x+\delta x/2} = T|_x + (dT/dx)|_x (\delta x/2)$, $T|_{x-\delta x/2} = T|_x - (dT/dx)|_x (\delta x/2)$. Therefore the net internal energy accumulation out − in : $(dT/dx)|_X C\rho u, \delta x\, \delta y\, \delta z = 0$ or,

$$\rho C u \frac{dT}{dx} = 0 \qquad (2.14)$$

2.6.5.3 Thermal Energy Conservation due to Diffusion

At $x = X$, as shown in Fig. 2.17 on page 22 we know the diffusion flux $Js^{-1}m^{-2})$, $q_X = -k\, dT/dx|_X$ (heat flow per unit area), at the control volume centre, in the x-direction is q_x, from eqn. 2.12. The area of the control volume surfaces normal to the x_1-axis is $\delta y\, \delta z$. The heat *flow* Js^{-1}, at the control volume centre, in the x-direction is $Aq_x = q_x \delta y\, \delta z$. The heat diffusion *out flow* Js^{-1} at $x + \delta x/2$, in the x-direction is $\delta y \delta z\, (q_x + (\delta x/2)(d(q_X)/dx))$. The heat diffusion *in flow*, at $x - \delta x/2$, in the x-direction is $\delta y \delta z\, (q_x - (\delta x/2)(d(q_X)/dx))$. The net heat diffusion flow, in the x-direction is $+\delta x \delta y \delta z\, (d(q_X)/dx)) = -\delta x \delta y \delta z k(d^2 T/dx^2)$.

2.6.5.4 1D Steady Flow Convection-Diffusion Equation

In the previous two sections we have shown we have two methods of transporting information across the surface: convection and diffusion. Therefore *for this problem* the energy conservation law is: $\rho C u(dT/dx)\delta x\, \delta y\, \delta z - \delta x\, \delta y\, \delta z\, k(d^2 T/dx^2) = 0$ or $u(dT/dx) = (k/\rho C)(d^2 T/dx^2)$. The coefficients of the 2nd order term are all material constants – assume $Z = (k/\rho C)$, therefore

$$u\frac{dT}{dx} = Z\frac{d^2T}{dx^2} \tag{2.15}$$

In our problem, x varies from 0 to 1, $T(x=0) = 0$, $T(x=1) = 1$, u and Z are constants (for Z best to think of only k can change). Fig. 2.25 shows three examples, (i) $u = 10$, $Z = 1$ (ii) $u = 100$, $Z = 1$ and (iii) $u = 10$, $Z = 5$. The first case is balanced convection and diffusion, the second convection dominates, and even though the diffusive gradient is very steep, the diffusive flux cannot complete with the convective power of the flow speed. The final example shows where the diffusion dominates. This example highlights how the relative strength of the diffusion and convection flux can define the temperature profile in our region of interest. This Convection-Diffusion balance principle is common to virtually all fluid transport problems.

2.6.5.5 1D Steady Flow Non-Dimensional Convection-Diffusion Equation

In what follows fluid physical properties (e.g. k) are normalised (e.g. k^*) by using a reference value that is constant throughout the domain (e.g. k_o). The same applies to spatial and temporal scales, and also differential quantities. So for instance in the dimensional equation $k = k^* k_o$, and so on for other variables. Differential operators must also be non-dimensionalised since $x^* = x/x_o$, $dx^*/dx = 1/x_o$, so $d/dx = (1/x_o)d/dx^*$. The idea behind this is that the magnitude of the variable is held in the dimensional term (u_o say) and the non-dimensional term (u^* say) is of order 10^0. For instance a trivial examples is if $u = 10m/s$, $u_o = 10m/s$ and $u^* = 1$. Or, if the flow varies between $100m/s$ and $130m/s$, $u_o = 100m/s$ and $1.3 > u^* > 1$. The real power of the technique however is when groups of these dimensional variables are made non-dimensional, these groups tell you how the equations (i.e. the physics) behave. In the present example we use this to *define* the relative strength of convective and diffusive flux, but is completely general, as shown when non-dimensioning the full governing equations in section 8.8 on page 115. This is a somewhat more intuitive way to use non-dimensional numbers. Non-dimensioning eqn 2.15 gives $[u_o T_o/x_o]\, u^*(dT^*/dx^*) = [C_o T_o/x_o^2]\, C^*(d^2T^*/dx^{*2})$ and ultimately $u^*(dT^*/dx^*) = [1/Pe_o](d^2T^*/dx^{*2})$. The Peclet Number defines the balance between Convective and Diffusive flux $1/Pe_o = [(k_o/(\rho_o C_o u_o x_o)] = [\mu_o/(u_o x_o \rho_o)]\,[k_o/(\mu_o C_o)] = 1/\text{Re}_o\,\text{Pr}_o$. When the Peclet Number is large, this tells us the transport mechanism is primarily convective. This is important physically, and also numerically, as discussed further in section 14.5.1 on page 152. The Peclet Number here is decomposed into the product of the Reynolds and Prandtl Numbers. The Prandtl number is the ratio of viscous to thermal diffusion flow, as discussed in section 2.4.5 on page 20. The Reynolds number has also been discussed previously in section 2.4.4 on page 19.

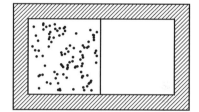

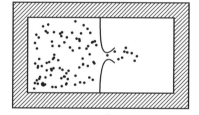

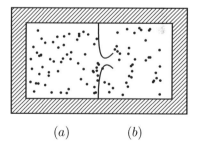

(a)   (b)

Figure 2.26: *Unrestrained Volume Expansion*

2.7 Definition of a Pure Single State Gaseous Fluid (Microscopic Description)

All materials are made up of atoms or molecules, and to *really* understand how materials respond to change, we ought to track each atom or molecule. We don't though, much of engineering thermofluids does not concern itself with microscopic (molecular) physics. Rather we define change in terms of macroscopic properties that are an average over many molecules. For instance density is the mass of a (very large) number of molecules in a volume. You need to be aware of the assumptions used to generate these macroscopic quantities (one assumption is the continuum assumption, see section 2.5.1) and to do this we need to examine how we obtain macroscopic quantities for a finite volume in terms of a population of molecules within it.

2.7.1   Kinetic Theory

We will do this for the simplest fluid possible, a low pressure gas composed of simple molecules. This analysis is a reasonable approximation to reality to many low pressure gas systems, for instance air at atmospheric conditions. As molecules become more complex and/or the gas pressure increases corrections need to be added to this description, as described in chapter 11 on page 136. For liquids, this analysis does not give accurate answers due to the strength of inter-molecular forces. An excellent example of this is the way in which viscosity changes with temperature. Kinetic Theory predicts that the viscosity goes up with temperature, which is what happens to gases, however liquid viscosities decrease as temperature rises.

2.7.2   Lord Kelvin and the One-Point Temperature Scale

The early work on gases was experimental and was useful in understanding how to measure heat, though at the time no-one had much idea what heat actually was. Boyle's law ($\sim$1662) found that as long as temperature is constant the product of absolute pressure and volume is constant when a gas is expanded, $pv = p/\rho = const.$

Charle's law ($\sim$1780) found that as long as the pressure stays constant, gas volume and temperature are directly proportional, $T/v = \rho T = const.$ Avagadro ($\sim$1811) found out a certain volume of gas at a certain temperature and pressure held a certain number of molecules. Clapeyron combined Charles's law with Boyle's law (1834) to produce a single statement which would become known as the ideal gas law $p/\rho T = const.$ Amontons ($\sim$1700) discovered that the pressure of a fixed mass of gas kept at a constant volume is proportional to the temperature. Amontons discovered this while building an "air thermometer". Fahrenheit's mercury thermometer defined a 180 point (degree) scale between the freezing point of a brine mixture and the boiling point of water. Similarly the Centigrade scale defines 100 points (degrees) between the freezing and boiling points of water. In these two point methods the measurement of temperature is relative to some arbitrary datum with an arbitrary number of intervals to another arbitrary datum. There is no physical meaning in these two point scales.

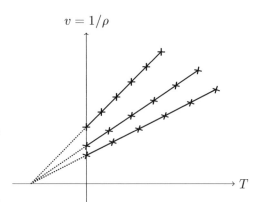

Figure 2.27: *Change in Specific Volume with Temperature for different gases*

Kelvin in collaboration with several others noticed that if you apply Charles's Law to several low pressure gases we obtain several different gradients on $v - T$ plot, as shown by Fig. 2.27 and noticed that (1) these are linear for a reasonable range of $v - T$ space, (2) if you extrapolate these back to lower and lower T you they all cross the $v = 0$ axis at the same point. This implied that all gases, at some fundamental temperature, have zero volume. The hypothesis was that if a gas volume is defined by the motion of gas molecules, and not the volume of the molecules themselves, then this temperature defines when molecules have zero motion, and therefore zero kinetic energy. If molecules have zero motion, then a gas comprising such molecules also has zero internal energy. Kelvin defined his reference temperature using this zero energy condition. It is a one point scale, and simply chooses the Centigrade degree as an arbitrary interval, hence the Kelvin scale is related to the Centigrade Scale by

$$T(K) = T\,(^{\circ}C) + 273.15.$$

(2.16)

Therefore the Kelvin temperature scale might be related directly to the gas internal energy, in other words it has a physical basis. Note: sometimes you have to use K scale temperatures, and sometimes it does not matter. It is best to always use K scale temperatures to calculate quantities correctly.

2.7.3 Relating Molecular Dynamics to Gas Pressure and Temperature

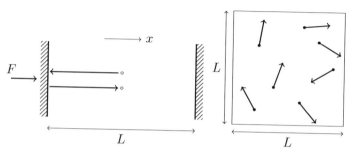

Figure 2.28: *A box containing some molecules moving in random directions (R) and one molecule moving normal to a wall (L)*

Here the aim is to relate macroscopic gas quantities such as pressure and temperature to microscopic properties such as the molecular speed and the number density. Consider an open cubic box of side L and volume V into which molecules of a simple low pressure gas at some T and P are allowed to pass in and out of. Then we close the box, so it is isolated from the environment, and wait till the system has settled down. Each of the N molecules has a mass m and velocity components u_x, u_y, u_z. Let us first examine a simple case as shown in Fig. 2.28. Let us take one molecule, moving in the x-direction only, and consider the impact on one wall of the box normal to that direction. The momentum (Ns) change of the particle under collision with LH wall = momentum before − momentum after = $-mu_x - (mu_x) = -2mu_x$. Therefore the force required to stop the wall moving is $2mu_x$. The number of times per second the wall is hit by this particle is $|u_x|/2L$. Therefore the momentum (kgm/s) change per second (N) on one wall = mu_x^2/L. Therefore the pressure on the wall, due to this one particle is $p = F/A = mu_x^2/LA = mu_x^2/V$.

In reality each particle is moving in three directions and $u^2 = u_x^2 + u_y^2 + u_z^2$, where u is the speed of the particle. If we sum over all the N particles $\sum_{n=1}^{N} u_n^2 = \sum_{n=1}^{N} \left(u_{x,n}^2 + u_{y,n}^2 + u_{z,n}^2 \right) = \sum_{n=1}^{N} u_{x,n}^2 + \sum_{n=1}^{N} u_{y,n}^2 + \sum_{n=1}^{N} u_{z,n}^2$. Since there is no preferred direction then $\sum_{n=1}^{N} u_{x,n}^2 = \sum_{n=1}^{N} u_{y,n}^2 = \sum_{n=1}^{N} u_{z,n}^2$ and so $\sum_{n=1}^{N} u_n^2 = 3 \sum_{n=1}^{N} u_{x,n}^2$. Finally we define the mean kinetic energy of the particle speeds as $\overline{u^2} = 1/N \sum_{n=1}^{N} u_n^2 = 3/N \sum_{n=1}^{N} u_x^2$. Remember the pressure on the wall, due to one particle is $p = mu_x^2/V$ so the pressure on one of the walls, from all the particles, moving in all directions is $p = (2N/3V)\left((1/2)m\overline{u^2} \right)$. So, we have the pressure–volume product equal to the mean kinetic energy of the N particles in our box, $pV = (2N/3)\left((1/2)m\overline{u^2} \right)$. We now have macroscopic properties on the LHS and microscopic properties on the RHS and the result looks a bit like Boyles Law (section 2.7.2).

We define the mean molecular kinetic energy proportional to temperature, what we are doing here is using a single *macroscopic* property, temperature to represent the integral effect of many molecular velocities, using our continuum assumption, $(1/2)m\overline{u^2} \alpha T$. Later, when discussing entropy on page 34 we refer to each velocity as a probability state. The proportionality constant is known as the Boltzmann constant, k_B (J/K) and relates molecular kinetic energy (J) to Kelvin (K). Thus the Kelvin temperature scale has fundamental meaning. $(1/2)m\overline{u^2} = (3/2)k_B T$ which, via $pV = (2N/3)\left((1/2)m\overline{u^2}\right)$ gives a form of the gas law $pV = Nk_B T$. It should be noted that what we have done is related *translational* kinetic energy of the molecules to the temperature via the Boltzmann constant. For more complex molecules rotational and vibrational energies also exist and hence kinetic theory gets progressively less accurate as the molecular complexity increases.

2.7.4 The Ideal Gas Law

This tells us the pressure–volume product is a function of T (i.e. the average kinetic energy of the mean particle, and the number of particles in the box). If the box volume and contents stays the same, the pressure rises linearly with T. This is exactly what Amontons discovered this while building an "air thermometer" in ~1700, but did not know why. Finally we may relate the Boltzmann constant to the Universal Gas Constant $(Jmol^{-1}K^{-1})$ using Avagadro's Number (mol^{-1}) $k_B = R_u/N_A$ and we recover the familiar molar gas law as follows: $pV = nR_u T$ here $n = N/N_A$ is the moles of gas present. To obtain the gas law in the more familiar mass basis, we start from the gas law per mole and multiply by the molar mass of the gas, $m_u = m/n$ to give $pV = m_u n(R_u/m_u)T = mRT$. Here R is the mass gas constant $(Jkg^{-1}K^{-1})$. This IS dependent on the type of gas since $R = R_u/m_u$. Finally, dividing both sides of the gas law by m gives the intensive form of the equation:

$$pv = \frac{p}{\rho} = RT \tag{2.17}$$

This is known as the ideal gas equation and is a very good approximation for low pressure gases. Gases that follow this law are known as ideal gases.

2.7.5 Gas Specific Heat at Constant Volume (Gas Internal Energy)

Considering our box of N molecules of mass m, we note it is a constant volume device. If we heated the walls and the molecules picked up (microscopic) kinetic energy over time the macroscopic energy measure (Kelvin temperature) would rise. *Internal* energy of a fluid is defined by the specific heat at constant volume, e.g. $E_u = (1/2)Nm\overline{u^2} = (3/2)Nk_B T = (3/2)nR_u T = (3/2)mRT = mC_v T$. As a specific quantity, $e_u = C_v T$. Therefore, $C_v = (3/2)R$. Formally, C_v is defined as the energy required to raise $1kg$ (not $1mol$) of gas by $1K$ at constant volume. $C_v = \partial e_u/\partial T|_v$. Note: the specific heat property is a material property. As we will see in section 2.9.7 on page 38 energy is a process property, something that changes a fluids state. In reality the specific heats are a function of temperature. An ideal gas that has constant specific heats is known as a perfect gas. With the exception of chapter 11 on page 136 only perfect gases are considered in this book.

2.7.6 Gas Specific Heat at Constant Pressure (Gas Enthalpy)

To introduce the constant pressure specific heat, we now imagine our box expands in one direction when the walls are heated and the molecules pick up speed (as above) to keep the pressure constant. Here the gas molecules are using up some of their energy to move the box wall out as well as moving faster (i.e. getting hotter). The energy expended in moving one face of the box out by δL is: $\delta W = F\delta L = pA\delta L = p\delta V$. From the gas law (equation 2.17), at constant pressure, $pV = mRT$, $\delta W = p\delta V = mR\delta T$. So, energy added at constant pressure, $E_h = mC_v\delta T + mR\delta T = mC_p\delta T$ or as a specific quantity, $e_h = C_p T$. Since $C_v = 3R/2$, $C_p = 5R/2$. This is the gas enthalpy, and is the sum of the internal energy of the gas and the energy required to increase the volume of the gas at constant pressure by a given temperature,

$$e_h = e_u + \frac{p}{\rho} \tag{2.18}$$

Formally, C_p is defined as the energy required to raise $1kg$ (not $1mol$) of gas by $1K$ at constant pressure. $C_p = \partial e_h/\partial T|_p$.

2.7.7 Enthalpy, Internal Energy, and the First Law of Thermodynamics

Our two specific heat definitions allow us to define the basic energy conservation equation, the First Law of Thermodynamics, stated without proof previously in section 2.3.1 on page 15. For a heat transfer at constant *volume*, the only thing that happens is that the gas temperature changes from state 1 to state 2, or $e_{u,2} - e_{u,1} = e_{u,12} = \int_{T_1}^{T_2} C_v \, dT = C_v \, (T_2 - T_1)$. When a heat transfer occurs at constant *pressure* the heat transferred always does two things. Some of the heat increases the temperature and some does some work on the environment. $e_{h,2} - e_{h,1} = e_{h,12} = \int_{T_1}^{T_2} C_p \, dT = C_p \, (T_2 - T_1)$. Relating these two definitions gives, $E_h = mC_v \, \delta T + mR \, \delta T = mC_p \, \delta T$, where the work done is $\delta W = p \delta V = mR \, \delta T$ or in specific form $\delta w = p \, \delta v = R \, \delta T$. In integral form, $W_{12} = p(V_2 - V_1) = mR(T_2 - T_1)$ and similarly for the specific form. Therefore $e_{h,12} = e_{u,12} + w_{12}$, since we can define specific work in terms of specific volume. So if we have a heat transfer (Q) into a mass (a system) at constant pressure.

$$Q_{12} - W_{12} = E_{u,12} \tag{2.19}$$

This is the First Law of Thermodynamics, in specific (per unit mass) form $q_{12} - w_{12} = e_{u,12}$, and is a simplification of the fundamental energy equation (2.9) on page 15.

2.7.8 Further Comments on the Specific Heats and Heat Transfer

For gases it also follows that the specific heats and the mass gas constant are related as follows: $C_p = C_v + R$ and another useful parameter is the ratio of Specific heats $\gamma = C_p/C_v$. If all gases obeyed kinetic theory then $\gamma = 5/3 \sim 1.66$ always. This is only true for simple inert gases such as Helium. Real gases such as air, oxygen and CO_2 have γ in the range 1.3-1.4 because these gases are molecules and have other energies (for instance vibrational) in additional to the translational (kinetic). Obviously, the distinction between C_v and C_p for incompressible materials is meaningless, there is only one specific heat.

Also a word on sign convention for the work and heat transfers enshrined in the first law. A heat transfer into the system (heating the gas up) is positive. A work transfer from the system to the environment is positive (a piston expanding a volume). This is the engineering convention, because heat engines generally take in heat and give out work.

Finally a comment about internal energy and enthalpy, and heat transfers. The change in the fluids temperature is *always* defined by the change in the internal energy. If the heating process is a constant volume heating process, then the heat transfer is the same as the internal energy change. If the heating process is a constant pressure process (ie some work and some heating is done) the heat transfer is defined by the enthalpy change of the gas and the temperature change by the internal energy change.

2.8 Entropy and the Incompleteness of the Gas State based on the First Law

First, let us remind ourselves what we have defined, because entropy is a very real and very important variable that is conceptually is difficult to visualise. We have some state variables that define the state of a gas. For a gas, P, V, T (extensive form) or P, v, T (intensive form) define the <u>state</u> of a gas. They are known as state variables. We have some universal constants, that apply to all gases. The Boltzmann constant k_B: relates (microscopic) molecular kinetic energy to (macroscopic) temperature. The Universal Gas Constant R_u: relates the (microscopic) number of molecules to the (macroscopic) energy they contain. We also have some material constants, that define the capacity of the gas to change state C_p, C_v, and from these R, γ.

The incompleteness of our knowledge is that our mathematical, molecular description of a gas does not fit with reality. The reality for our simple gas in a box (defined in section 2.7.3 on 31) is that it will tend to expand into any available space and also that a hot gases will tend to transfer heat to a cooler gas if the two were mixed. However, equation 2.19 on page 33 does not define this observed *directionality* of heat flow. We need more information, another constraint, another *law*. Unfortunately, this law cannot be written exactly using a fixed mass or number of molecules.

Maxwell realised that this process could only be satisfied in the statistical sense. Boltzmann and Clausius realised that entropy either stays constant or increases over a cycle. Bizarrely, it is not a conserved quantity. So, key differences between entropy and energy are: (1) Energy is well behaved conserved quantity we can develop precise conservation laws for, (2) entropy cannot be defined precisely, is not conserved and at best, stays constant. There are far reaching implications of entropy generation in engineering which include: (1) the inability to convert all heat energy to work and (2) the loss of useful energy in fluid systems due to friction and other irreversible losses.

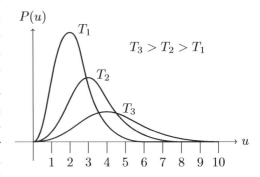

Figure 2.29: *The probability of a certain molecule having a certain speed for different gas temperatures*

2.8.1   Definition of Molecular (Microscopic) Entropy

The reason this is important is that energy has a quality as well as a quantity measure. If we mix two gases together, of two different temperatures the total energy does not change, but we will have lost some useful energy. We could have put some machine between the hot and the cold gases and used that heat flow to do something useful. But we just mixed the gases, and now we can never unmix them to do anything useful. This loss of useful energy even though the total energy present does not change is one way to visualise entropy.

Another problem with understanding what entropy is its inexactness. Maxwell explained the inexactness of entropy in terms of an example, "Maxwells Daemon". He imagined two compartments with a tiny hole connecting them containing gas at the same temperature. Therefore each compartment has a population of molecules with the same speed distribution. He then imagined a daemon doorkeeper, letting through the fast molecules into one compartment, and slow ones through into the other. Over time, one compartment would heat up and one would cool down. He reasoned you do not actually need the doorkeeper, over short times this might actually happen, but over long times it never will.

The key point is you can never prove this by considering a single or a few particles. You have to consider many particles, over a sufficiently long time and claim that reality is actually only very very likely. Hardly comforting is it?

Figure 2.30: *Entropy Change due to the increase in the number of position "'states'".*

Recall that we defined temperature as a variable that represents the mean kinetic energy of the molecules in our box. Higher temperature means more kinetic energy present, and the faster our molecules are moving, on average. It turns out that we can work out exactly the probable speed distribution of the molecules in the box, solely from the mass of the molecule, the temperature and it is known as the Boltzmann Distribution. It is sketched for three different gas temperatures in Fig. 2.29, and shows the probability of the speed of a gas molecule on the y-axis in terms of speed "'states"' on the x-axis. Think of these states as "'... very slow, slow, fast, very fast...etc"'. The Boltzmann Distribution has the following properties. (1) It has a high velocity tail: there is also a low probability of very high speed molecules present. (2) Higher energy distributions (hotter gases) have a wider range of possible velocities (more "states"). (3) Because of this the probability that any one molecule will have a certain velocity is lower for higher energy distributions.

It is the *range of possible states* that is the key to understanding entropy at a microscopic level. As shown in Fig. 2.29 a *state* is a quantised speed interval. Let us say our quantised interval is $10m/s$, the actual number is irrelevant as long as it is small relative to the difference between the fastest and the slowest molecules.

Now look at Fig. 2.29 and imagine counting the number of speed states (intervals) the gas molecules at the two different temperatures can inhabit. You can see 10 speed states and for the $T = T_3$ distribution, molecules can have any one of ten quantised speed states. The range of states for $T = T_3$ is ten. Purely visually you can see that the lower temperature gas has a fewer number of possible speed states than the higher temperature gas. The more possible states (the more possible "disorder") the bigger the gas entropy. Lower temperature gases have lower entropy than higher temperature gases.

Entropy change can also be defined "range of possible states" using position instead of speed as outlined above. An increase in entropy comes from volume change in an insulated box, shown in Fig. 2.30. Consider molecules in one half of the box at $t = 0$, where the other half is a vacuum. The partition is removed and the molecules over time $t > 0$ fill the box. They are very very unlikely to ever go back, we can consider the process is irreversible. How has the state of the gas changed? The gas has expanded against a vacuum so no work has been done. The box is insulated, so there is no heat transfer and from equation 2.19 on page 33 the molecules have lost no internal energy. Therefore the kinetic energy of the molecules has not changed. The speed distribution and therefore the temperature has not changed. The range of speed states is unchanged. But, the macroscopic quality of the energy has however has degraded, because of the loss of pressure. In terms of microscopic "range of possible states" each molecule has more position states when the partition is removed. If we divide each section into (arbitrarily) 9 areas, at $t = 0$ the probability of finding a molecule in any one area is 1 in 9, after it is 1 in 18. This increases the molecular entropy. Again the number of possible states chosen is irrelevant, what is important is the relative change, here a factor of 2 more. For our low pressure simple gas system we have shown that the microscopic entropy can change even when the energy of the system stays constant, and for our simple system there are two reasons. (1) When gases of two different temperatures mix in the same volume and (2) when a gas expands in an unresisted manner to increase volume.

A definition of microscopic (and statistical) entropy is $s = k_B \ln N_S$ where N_S is the number of possible states. This explains the above, but is inexact for s because it depends on how big each state interval is (or in other words, how many there are). What we can do is state what the entropy *change* is exactly for instance, the entropy change due to the volume change is $s_{12} = k_B \ln 2$. In section 2.9.9 on page 39 entropy change for real systems is introduced, which uses a different definition, however it is shown in section 2.9.13.3 on page 43 that these two definitions are compatible.

2.8.2 Third Law of Thermodynamics

The definition of statistical entropy also gives us another fundamental reference point for gases. As the temperature of the gas tends to absolute zero on the Kelvin scale then so does the entropy. This can be understood in terms of the mean thermal speed of the molecules all tending to a single zero state. Then, the molecules all have the same (zero) speed, and hence all have the same state. In other words the probability of the molecules having zero speed is 1. Likewise the gas theoretically occupies zero volume, and therefore the position of the molecules are all known, therefore the number of possible position states a molecule might have is also 1.

Therefore in velocity-position space, there is only one molecular state at absolute zero, therefore $N_s = 1$ and $s = k_B \ln N_s = 0$. This is sometimes known as the Third Law of Thermodynamics. It gives a further physical basis into the Kelvin temperature scale defined by eqn. 2.16 on page 31.

2.9 Energy and Entropy Conversion by Changing a Fluids State

2.9.1 Internal Equilibrium, Two Property Rule

If a system is isolated from its environment and nothing is changing the system (a gas in this case) is said to be in internal equilibrium. The rigid (no work transfer) insulated (no heat transfer) box shown in Fig. 2.26 contains two compartments (a) containing a gas and (b) a vacuum, separated by a diaphragm. The system is in internal equilibrium in section (a). Nothing is in section (b). The diaphragm breaks and gas properties (p,T) vary in space and time, and the system is not in internal equilibrium. Finally the system is again in a new state of internal equilibrium. Note there is a state of dynamic equilibrium between the two sections at the final time. Internal equilibrium is thus a state whereby a fluid property may be defined by a single value for that system. An important assumption can be made when a system is in equilibrium, in that the two property rule is valid. This states that:

- The state of a simple compressible system is fixed completely by values of any two of its intensive properties, provided they are independent of each other.

For example if we know the gas pressure and temperature, we can work the density (and the entropy).

2.9.2 Forms of Equilibrium between System and Environment

Relaxing the constraint that the system is separated from the environment thermally and work is being transferred between the system and the environment provides for further types of equilibrium conditions. In Fig. 2.31 we consider initially a gas contained in a cylinder with a stop on the cylinder. The cylinder gas is in internal equilibrium since nothing is in motion and no thermal contact to the environment. If the cylinder insulation is removed and the stop taken away, then both the piston starts moving and also heat transfers through the cylinder walls. Neither internal, mechanical nor thermal equilibrium is valid here. The piston motion slows and eventually stops as does the net heat flow across the cylinder wall. The system is in mechanical equilibrium (no change in the piston position in time) thermal equilibrium (zero *net* heat transfer across the cylinder wall) and internal equilibrium (uniform conditions in the system).

2.9.3 Zeroth Law of Thermodynamics

A principle that was implicitly assumed when developing later laws of thermodynamics is one of implicit equilibrium across several systems in the same environment. This is known as the Zeroth Law of Thermodynamics, "If system A is in thermal equilibrium with the environment and system B is in thermal equilibrium with the environment then system A and B are also in thermal equilibrium".

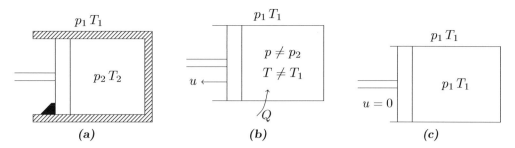

Figure 2.31: *Transition between Equilibrium States*

2.9.4 Initial and Final states: Process and Path

Generally we have an initial state and a final state, some method to get between states and the route taken. These initial and final states are always in a state of internal equilibrium, where no further change occurs. Therefore the two property rule applies to these states. Because of the two property rule, a position on a 2D state plot defines *completely* the state of the fluid. Let us say our initial state is cold water and our final state is boiling water. The change by which we get from initial to the final state is known as the process and the series of states the system passes through to get to the final state is known as the path. In our example the process is heat addition and the path could be the plot of temperature versus pressure. To specify the process (and define the path) the initial and final states and the transfers across the system boundary must be defined. Unless we make a further assumption we generally do not know the state of the system between the initial and final states.

2.9.5 Quasi-Equilibrium Processes

Non-equilibrium processes are very difficult to analyse, and in this book we will be restricted to processes that go from initial to final states under a quasi-equilibrium assumption. This states that that the system, moving along a process path from an initial to a final equilibrium state does so slowly enough to ensure the internal equilibrium is maintained. The key advantage is we can apply the two property rule all the way along the process path. This is often not such a bad assumption, for example during the compression of a gas in an engine cylinder the assumption is that pressure and temperature remain spatially uniform in the cylinder at all times. For an engine running at 4000rpm = 76rps, one compression cycle takes ~0.015 sec. The pressure information travels at the speed of sound ~340m/s and the cylinder stroke is about 0.1m, so pressure information takes 0.0003 sec to get from the piston to the cylinder head. Since the pressure information travels much faster than the piston, this suggests that the compression can be approximated as a quasi-equilibrium process.

2.9.6 Reversible and Irreversible Processes

In addition to the quasi-equilibrium assumption of a process we also need to consider whether the process is reversible or not. A reversible process will be able to go from an initial state to a final state and then back to the initial state using the same process. A reversible quasi-equilibrium work process is the slow compression of a gas in an insulated and frictionless piston-cylinder. The gas can expand and return to the original state. An irreversible quasi-equilibrium work process is the slow compression of a gas in an insulated piston-cylinder with friction present. The gas can expand but will not be able to return to the original state under the same process because of some energy loss. A reversible quasi-equilibrium heat transfer process is the very slow expansion of very large and thin walled frictionless piston-cylinder at constant temperature. As the gas expands the heat transfer into the cylinder occurs without a temperature difference. Therefore the process is reversible. An irreversible quasi-equilibrium heat transfer process is the constant volume heating of a cold gas. If a temperature difference exists between the environment and the system then heat cannot flow out of the system and therefore the process is not reversible.

It is irreversible change that is the major engineering problem. Practically, it cannot be avoided it can only be minimised. This requires understanding of the fundamental physics and the design of technology that minimises irreversible entropy generation. A good engineer is an entropy minimiser. Examples of irreversible entropy generation in thermofluids:

- Friction (Shear Stresses) in fluid flow – leads to pressure drops in pipes, drag on objects (chapter 5 on page 75).
- Turbulent flow – much larger pressure drops in pipes, drag on objects (chapter 13 on page 144).
- Heat transfer across a temperature difference (chapter 12 on page 141).
- Unrestrained expansion (section 2.9.13.3 on page 43)
- Shock waves in compressible fluids (chapter 10 on page 127).
- Mixing of fluids (section 2.9.13.2 on page 42).
- Chemical Reaction/Combustion (not covered here)

2.9.7 $p - V$ diagrams of process paths: Heat and Work as Process Properties

Providing our process is quasi-equilibrium then at all points along the path, the state of the fluid between the initial and final states is *fully* defined by the two property rule. $p - V$ diagrams are very useful for systems involving volume change of the working fluid as a function of pressure: *i.e.* *displacement work*: work moving the system boundary. Note however if work does follow a quasi-equilibrium path the process should be "slow". The term used most is "fully resisted" and has important consequences for entropy generation as noted in section 2.8.1 on page 34. Therefore, since $\delta V = A \, \delta x$, $\delta W = F \, \delta x = pA \, \delta x = p \, \delta V$, $\int dW = \int p \, dV$ (if the process is reversible). Note work is *not* a <u>state</u> property (of the fluid), it is a <u>path</u> property (of the <u>process</u>). In other words work does not define a fluid state but the *transition* between an initial and a final state. Therefore from now on we represent work as Work (initial state to final state), or in this case W_{12}, defining the work done changing the gas state from state 1 to state 2. Since heat and work are convertible heat transfer is also a process variable. State properties are P, V, T, S. Work and heat transfers across system boundaries are process, *not* state properties.

2.9.8   Work and Energy Change in Processes

The four processes shown in Fig. 2.32 are very important in thermofluids, and in what follows the First Law (equation 2.19, page 33) is employed to relate work, heat transfer and internal energy (temperature) changes in a gas. The gas law (equation 2.17, page 2.17) and internal equilibrium at initial (state 1) the final (state n, where n=2-5) process states defines $p_1 V_1 / T_1 = p_n V_n / T_n$ in all cases. As noted in section 2.7.8 on page 33 for incompressible fluids $C_v = C_p$, hence isochoric and isobaric heating are equivalent.

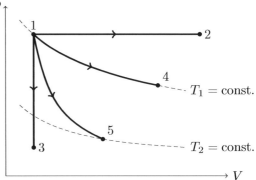

2.9.8.1   Isochoric (Constant Volume) Heating/Cooling

As shown in Fig. 2.32 from state 1 to state 3 the gas is cooled, decreasing the pressure, in a rigid container. The inverse occurs during the (usually very short) combustion period of an spark ignition internal combustion engine

Figure 2.32: *p-v change for key heat and work transfer processes*

cycle. Since $V_1 = V_3$, from the gas law $p/T = const$ defines the constant volume process. From the First Law $W_{13} = 0$ since the volume is constant and thus all of the heat transfer contributes to internal energy (temperature) change. $Q_{13} = mC_v (T_3 - T_1) = E_{u,13}$. Since the heat transfer occurs over the temperature range because the gas temperature varies while the external cooling temperature is fixed the process is not reversible.

2.9.8.2 Isobaric (Constant Pressure) Heating/Cooling

For constant pressure expansion the gas is heated, increasing the volume from V_1 to V_2 as shown in Fig. 2.32. The temperature rises to maintain the pressure as volume increases. This occurs in piston expansion and also a range of constant flow process equipment requiring heat transfer, for instance heat exchangers, boilers. From the gas law $p_1 = p_2$ and thus $V/T = const$. The work done in extensive and intensive form is $W_{12} = \int dW = \int p\,dV = p(V_2 - V_1)$, $w_{12} = \int dw = \int p\,dv = p(v_2 - v_1)$. Energy raises the internal energy and does displacement work. $Q_{12} = W_{12} + E_{u,12} = mC_p\,(T_2 - T_1) = p(V_2 - V_1) + mC_v\,(T_2 - T_1)$. This is the enthalpy change $E_{h,12} = E_{u,12} + pV$ or in intensive form: $e_{h,12} = e_{u,12} + p/\rho$. The heat transfer occurs over the temperature range, since the system temperature is changing while the environment temperature stays fixed, so the process is not reversible. If friction is present on the piston more irreversibility occurs.

2.9.8.3 Isothermal (Constant Temperature) Expansion/Compression

As shown in Fig. 2.32 during an expansion from state 1 to state 4 heat is added to maintain the temperature of the working fluid at T_1. The requirement for isothermal conditions is not usually useful practically. From the gas law here $T_1 = T_4$ and thus $pV = const$ defines the process. The work transfer is $W_{14} = \int \delta W = \int p\,\delta V = p_1 V_1 \int \delta V/V = p_1 V_1 \ln(V_4/V_1)$. And since there is zero change in internal energy then $Q_{14} = W_{14}$. Since all heat transfer occurs at the isotherm temperature: the process is reversible if frictionless. Note the condition that the heat transfer does not occur over a temperature difference requires an extremely large heat transfer area and an extremely slow expansion/compression process.

2.9.8.4 Adiabatic (Zero Heat Transfer) Expansion/Compression

Expansion occurs from state 1 to state 5 as shown in Fig. 2.32, in an insulated container therefore all of the work donated from the fluid serves to cool the fluid, $p\delta V = -mC_v\delta T$. This may be manipulated to give the process definition, $pV^\gamma = const$. It is a good approximation for the power and compression strokes in internal combustion engines. The work transfer is $W_{15} = \int dW = \int p\,dV = p_1 V_1^\gamma \int dV/V^\gamma$. Zero heat transfer therefore $Q_{15} = 0$ and $E_{u,15} = -W_{15}$. Because of the zero heat transfer the process is reversible if the piston motion is frictionless.

2.9.9 Macroscopic Entropy

Figure 2.33 shows we have two points defined on a $p - V$ diagram, 1 (initial state) and 3 (final state). First let us look at the process path $1 \rightarrow 2 \rightarrow 3$. This is a reversible isothermal expansion followed by a reversible adiabatic expansion. During $1 \rightarrow 2$. Isothermal expansion occurs and $Q_{12} = W_{12}$. Note also that the heat provided to the gas is all provided at a single temperature, $T_1 = T_2$ and thus the internal energy does not change. During process $2 \rightarrow 3$ an adiabatic expansion occurs and $E_{u,23} = W_{23}$. So in summary during the process $1 \rightarrow 2 \rightarrow 3$ Q_{12} has been supplied to the gas and the internal energy has changed.

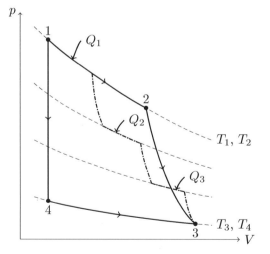

Figure 2.33: *Several Process Paths Producing the Same Change in Entropy*

Now let us look at the process path $1 \rightarrow 4 \rightarrow 3$, which has the same initial and final states. This is a reversible adiabatic expansion followed by a reversible isothermal expansion. Process $1 \rightarrow 4$ is the adiabatic expansion and $E_{u,14} = W_{14}$ while for process $4 \rightarrow 2$, an isothermal expansion process the heat transfer is $Q_{43} = W_{43}$. Note also that the heat provided to the gas is all provided at a single temperature, $T_4 = T_3$ and thus the internal energy does not change. In summary during the process $1 \rightarrow 4 \rightarrow 3$ Q_{43} has been supplied to the gas and the internal energy has changed.

Paths $1 \rightarrow 2 \rightarrow 3$ and $1 \rightarrow 4 \rightarrow 3$ are both reversible, and start and end at the same points. On paths $1 \rightarrow 2 \rightarrow 3$ and $1 \rightarrow 4 \rightarrow 3$:

- The heat transfers are Q_{43} and Q_{12} are different since T_4, $T_3 \neq T_1$, T_2.
- The work transfers are W_{14} and W_{23} are also different.
- The internal energy change between the initial and final states is the same for either path.
- Because the initial and final states are the same irrespective of the path, then the entropy (a state variable) must also change by a defined amount.

So although the initial and final states are the same the work and energy transfers are different on two different paths. The sum of the energy and work transfer are the same, since the total internal energy change is the same for both paths. This is simply a restatement of the first law. One other thing is the same for both paths the quantity of heat added divided by the temperature at which it is added, $Q_{43}/T_3 = Q_{12}/T_1 = S_{43} = S_{12}$. Since process $1 \rightarrow 4$ and $2 \rightarrow 3$ on the two paths generate no entropy since they are isentropic then $S_{13} = S_{12} = S_{43}$.

Note we could choose a more complicated route consisting of any series of adiabatic and isothermal expansions as shown in Fig. 2.33 and the total entropy change between states 1 and 3 would be the sum of the heat transfers during the isothermal processes. Here n is the index of the N isothermal processes.

$$S_{13} = \sum_{n=1}^{N} \frac{Q_n}{T_n} = \int_{1}^{3} \frac{dQ}{T}. \tag{2.20}$$

Because the entropy change between the initial and final states 1 and 3, S_{13} is the same irrespective of the reversible path taken, the value of entropy at the initial and final states must be the same. Because the variable is independent of the reversible path taken, it must be a state property, and thus there are 4 state properties, S, P, V, T that define a gas. If this is true we must be able to define relationships between them because of the two property rule and these are the $T\,dS$ equations.

2.9.10   The $T\,dS$ Equations

Total specific internal energy of a gas is $e_u = C_v T$ if T is in Kelvin. If the volume is held constant, then all the incoming heat transfer q goes to raising the temperature (internal energy). Now suppose then we add the same energy δq but this time allow the volume to change to maintain constant pressure then $\delta q = C_p \delta T = C_v \delta T + p\,\delta v$. Dividing by T, using the gas law, $\delta q/T = C_v(\delta T/T) + R(\delta v/v)$. Therefore, integrating, $\int dq/T = C_v \int dT/T + R \int dv/v$ we arrive at the $T\,dS$ equations, which comes in two forms and are known as the *1st* and *2nd* TdS equations respectively.

$$s_{12} = C_v \ln \frac{T_2}{T_1} + R \ln \frac{v_2}{v_1},\, s_{12} = C_p \ln \frac{T_2}{T_1} - R \ln \frac{p_2}{p_1} \tag{2.21}$$

This is the intensive form – entropy here has units of $Jkg^{-1}K^{-1}$. Note that we used the entropy definition and the 1st Law and the enthalpy definition to derive the $T\,dS$ equations: They apply to both reversible and irreversible processes, and not just the reversible example given.

2.9.11   Entropy Change in Processes

We now use the $T - dS$ equations to examine the entropy changes in the four key processes defined in section 2.9.8 on page 38. These are sketched in Fig. 2.34. Of these isobaric and isochoric processes cannot be reversible because the heat transfer between system and environment occurs over a temperature difference: They are not isentropic because of the heat transfer *and* because of the irreversibility. The isothermal process is reversible because the heat transfer occurs over a negligible temperature difference. However because of the heat transfer the process cannot be isentropic. The <u>adiabatic</u> process is <u>isentropic</u> if it is <u>reversible</u>, by virtue of the entropy definition (equation 2.20 on page 40).

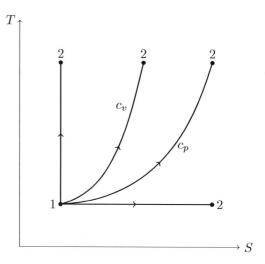

Figure 2.34: *T-s change for key heat and work transfer processes*

2.9.11.1   Isochoric (Constant Volume Heating/Cooling)

The entropy change here is due to an increase in the possible velocity states when the temperature rises. Using the first $T\,dS$ equation $S_{12} = mC_v \ln T_2/T_1$.

2.9.11.2   Isobaric (Constant Pressure Expansion/Compression)

Entropy increases due more than the constant volume process due to an increase in the possible position states *and* the possible velocity states. Using the 2nd $T\,dS$ equation: $S_{12} = mC_p \ln T_2/T_1$.

2.9.11.3   Isothermal (Constant Temperature Expansion/Compression)

Using the first $T\,dS$ equation $S_{12} = mR \ln V_2/V_1 = Q_{12}/T_1$. Note this reflects the result of equation 2.20 on page 40.

2.9.11.4   Adiabatic (Zero Heat Transfer Expansion/Compression)

Zero heat transfer gives zero entropy change if the process is also frictionless. Then the process is said to be <u>isentropic</u>.

2.9.12   Reversible $p - V$ and $T - S$ Diagrams

For reversible processes integrals in $p - V$ and $T - S$ diagrams have direct physical interpretation. $W_{12} = \int dW = \int p\,dV$. Note $W > 0$ for $dV > 0$ since $p > 0$ always $Q_{34} = \int dQ = \int T\,dS$. Note $Q > 0$ for $dS > 0$ since $T > 0$ always. Here we are examining changes in P, V, T, S for a single process: P, V, T, S are the state variables of our working fluid in our system. These changes are due to Q and W transfers between our system and the environment defined by the first law. Changes in P, V, T, S for any single process of the system may be > 0 or < 0 because W, Q change likewise. This conflicts with the standard knowledge, that entropy can only increase. In section 3.3 on page 45 we include the environment into this budget when examining process sequences (cycles) and later show that then change in $S \geq 0$ only is

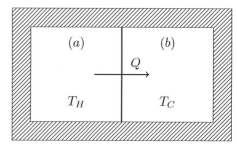

Figure 2.35: *Heat Flow Between Two Chambers*

possible.

The characteristics of the 4 key processes covered in section 2.9.8 on page 38 for energy change and reversibility, and section 2.9.11 on page 41 for entropy change are given below in table 2.2. This table is an excellent way to categorise these four key processes and to understand how work, heat transfer and internal energy are exchanged (as defined by the First Law, equation 2.19 on page 33) are related for each. The best approach for reconstructing the table is to first fill in the zero entries, from the process definitions. The First Law then tells you in most cases what then is equal to what. It also emphasises that reversible processes are not necessarily isentropic.

Process	W_{12}	Q_{12}	$E_{u,12}$	Reversible?	S_{12}
Isochoric $V_{12} = 0$	0	$mC_v(T_2 - T_1)$	$mC_v(T_2 - T_1)$	no	$mC_v ln(T_2/T_1)$
Isobaric $p_{12} = 0$	$p(V_2 - V_1)$	$mC_p(T_2 - T_1)$	$mC_v(T_2 - T_1)$	no	$mC_p ln(T_2/T_1)$
Isothermal $T_{12} = 0$	$\int p\,dV$	$\int p\,dV$	0	yes	$mRln(V_2/V_1)$ or Q_{12}/T_1
Adiabatic $Q_{12} = 0$	$mC_v(T_2 - T_1)$	0	$mC_v(T_2 - T_1)$	yes	0

Table 2.2: *Summary of Ideal Gas Process Characteristics*

2.9.13 Entropy as an Energy Quality Measure

Entropy is a difficult variable on conceptually understand because it is not measured directly but rather from measurements of the other state variables and then derived from the $T\,dS$ equations. It is however very important and the following examples outline in the simplest possible terms how entropy changes can define the useful energy lost from a system, even if there is no overall change in the energy present in that system.

2.9.13.1 Heat Flow Between Two Chambers

In Fig. 2.35 we have an insulated rigid chamber with a partition separating two equal sub-volumes. At state 1 the fluid in one sub-volume is hot, at temperature T_H and the other sub-volume is cold at a temperature T_C. At state 2 (some *time* later) both sub-volumes have the same temperature due to heat transfer through the partition. Between the initial and final states note that : No work transfer has occurred. No heat transfer into the system has occurred. Therefore from the first law the internal energy in the chamber has not changed. A heat transfer has occurred between sub-volumes and if we consider the each sub-volume a separate system (a and b) then the initial reversible entropy change in each system is $S_{b,12} = Q/T_C$, $S_{a,12} = -Q/T_H$, giving $S_{12} = S_{a,12} + S_{b,12} = -Q/T_H + Q/T_C > 0$. The entropy of the rigid container has increased. In other words the energy quality has decreased. We could in theory have placed some machine to make use of the energy transfer. Because we did not we have lost the use of that energy transfer forever.

2.9.13.2 Pressure Leakage Between Two Chambers

Imagine we have the same rigid container and this time, as shown in Fig. 2.35 the initial state 1 is the pressure in the two sub-volumes is different ($p_a = 2p_b$ say). The pressure then equalises very slowly over time, such that at the final state 2 the two pressures are equal. Between the initial and final states: No work transfer has occurred (no volume change). No heat transfer into the system has occurred.

Therefore from the first law the internal energy in the chamber has not changed. A mass flow has occurred between sub-volumes and if we consider the each sub-volume a separate system (a and b) then the entropy change in each system is characterised by the pressure change $p_a V = m_a R T$, $p_b V = m_b R T$, $m_a = 2m_b$, $p_2 2V = (m_a + m_b) R T$, $m_a = 2m_b$, $p_2 = 3m_a R T / 4V$, $p_2/p_a = 3/4$. The final pressure goes down with respect to p_a, $m_b = m_a/2$, $p_2 = 3m_b R T / 2V$, $p_2/p_b = 3/2$. Final pressure goes up with respect to p_b $S_{b,12} = -m_b R \ln (p_2/p_b)$, $S_{12} = S_{a,12} + S_{b,12} = -R m_b [2 \ln (p_2/p_a) + \ln (p_2/p_b)] > 0$, $S_{a,12} = -2m_b R \ln (p_2/p_a)$. Again entropy increases, energy quality has decreased, we have lost some useful energy forever.

2.9.13.3 Unrestrained volume expansion

A more subtle reason for an increase in entropy comes from volume change in an insulated box as shown in Fig. 2.30 on page 34. Initially we consider molecules in one half of box, where the other half is a empty vacuum. The partition is removed and the molecules over time fill the box. They are very very unlikely to ever go back – the process is irreversible. How has the state of the gas changed? The gas has expanded against a vacuum so no work has been done. The box is insulated, there is no heat transfer and the molecules have lost no internal energy. Therefore the energy of the molecules has not changed. The macroscopic quality of the energy has however has degraded because of the loss of pressure. Here we use the $T \, dS$ equations: $S_{12} = m R \ln V_2/V_1 = m R \ln 2$. Again entropy increases, energy quality has decreased, we have lost some useful energy forever. Recall $k_B = R_u/N_A$ and $m_u = m/n$, so $S_{12} = m R \ln 2 = N/N_A R_u \ln 2 = N k_B \ln 2 > 0$. Recall from section 2.8.1 on page 34 the molecular entropy definition was $S = k_B \ln N_S$ so clearly we have a correspondence between our microscopic entropy definition and our macroscopic entropy change. In this system, each of the N molecules have picked up $k_B \ln 2$ of entropy by increasing the number of possible states by $\ln 2$.

2.10 Key Points

We are now ready to examine thermofluid operations and understand how they operate based on the unit operations we have covered in this chapter, namely.

- A mathematical toolbox, allowing us to describe scalar and vector information and how it changes in space and time.
- Fundamental conservation laws, based on a fixed mass and how to convert them into forms based on a fixed volume.
- Dimensional variables and dimensionless groups.
- Material properties and characteristics of a fluid.
- How fluid information can be convected (by the flow) and diffused (by molecular interactions).
- Molecular motion and the link to the Kelvin temperature scale, ideal gas properties and relationships.
- Quasi-equilibrium processes, macroscopic entropy, and processes used for work and energy transfer.

3 | Conservation of Energy for a Fixed Mass

A thermofluid process (in other words some machine that either moves through a fluid, or has fluid moving through it) generally involves fluid motion, and therefore <u>convective</u> transport of momentum or energy, as discussed in section 2.6.5.2 on page 28. Also because of molecular interactions, diffusion of mass, momentum and energy is also present, as discussed in section 2.6.5.3 on page 28. Before we get onto this general case, in the next chapter, we examine thermofluid processes where there is no transport due to bulk motion present. Because there is no motion, there is no mass or momentum transfer, so we are considering only energy transfers for a fixed mass. So, what we are really doing is using a <u>system</u> basis to describe energy conservation due to work and diffusive heat transfers across the system boundary. This also requires the second law of thermodynamics and indeed, this chapter is really an analysis of this conceptually difficult aspect of thermofluids.

3.1 A Cycle of Processes

A <u>cycle</u> is a sequence of <u>processes</u> that start and end with the fluid in the same state, this has already been introduced briefly in section 2.2.7 on page 12. Therefore via the <u>two property rule</u> (section 2.9.1 on page 36) in a 2D plot of any two state variables, the start and end point of the cycle will share the same <u>state</u> co-ordinates. In the simple two process cycle has been shown previously in Fig. 2.9 on page 13, W_{12} is +ve (V increases) and W_{21} is -ve. As noted in section 2.9.8 on page 38 the net work is the area enclosed by the cycle and is clearly > 0 and for clockwise cycles this is always the case. The same is true on a $T-s$ diagram, where the net heat transfer is defined by the area enclosed on that cycle. The final important point is since the initial and the

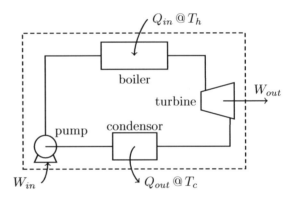

Figure 3.1: *A viable engine producing net work by taking in thermal energy at a high temperature and rejecting it at a lower temperature*

final points on the cycle are the same, the total internal energy change around the cycle is zero. This leads to the First Law of Thermodynamics for cycles to be the sum of all the heat and work transfers over the cycle,

$$\sum Q - \sum W = 0. \tag{3.1}$$

From this, we see that for a cycle $W_{net} = Q_{net}$ hence the area enclosed by a cycle on a $p-v$ and $T-s$ diagrams, W_{net} and Q_{net} are equivalent. In section 2.9.8 on page 38 we considered the energy and in section 2.9.11 on page 41 entropy changes in single processes. We showed that the energy and entropy change for a single process could be positive or negative. We were not concerned with the impact of dumping heat to the <u>environment</u>. As we shall see, it is crucial for good efficiency, and restricts the entropy change over a cycle to be positive or at best zero.

3.2 Heat Engine Fundamental Characteristics

A heat engine is any device that converts some of the heat energy into useful work and they all have the following characteristics:

- They receive heat energy from a high temperature source.
- They produce some work.
- They reject heat energy to a low temperature source.
- They operate on a cycle.

In the example shown in Fig. 3.1 all of these characteristics are present, and note we have to reject heat to a low temperature source because we have to get back to the initial state on the cycle. More specifically the fluid is a system, a fixed mass going around the cycle from process to process, changing state as it goes. The work and energy transfers are between that fixed mass of fluid (the system) and the environment. Obvious examples are internal combustion engines, steam and gas turbines. Less obvious examples are refrigerators and heat pumps, which are known as reversed heat engines and are discussed in section 3.7 on page 47.

3.3   Second Law of Thermodynamics: Kelvin-Planck Statement

As noted in section 2.2.7 on page 12 heat engines run on cycles that seem to require us to reject (waste) energy to a heat sink. This leads to several consequences enshrined in one of the more useful interpretations of the 2nd Law: The Kelvin-Planck statement.

It is impossible for any heat engine to receive heat from a single thermal reservoir and produce an equivalent amount of work.

Or

A cyclically operating engine must reject heat to a low temperature 'sink' as well as receive heat from a high temperature source.

Which means . . .

no heat engine can be 100% efficient.

This is because conceptually, efficiency $\equiv$ useful work out / energy in $=$ W_{net}/Q_{in}. As we shall see in section 3.8 on page 47, even the most efficient heat engine possible, the Carnot engine, can never be 100% efficient. Understanding why at a conceptual level is an indicator of understanding thermofluids.

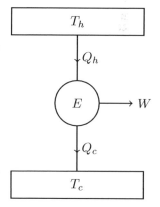

3.4   Thermal Reservoirs

Figure 3.1 shows a heat engine with 4 processes, a boiler, a turbine, a condenser and a pump. Instead of examining heat and work transfers for each process on the cycle, we can consider the *net* work to/from the system and the heat transfers to the hot and cold reservoirs. This is represented in Figure 3.1 by the transfers across the dashed line boundary. In this way instead of concerning ourselves with the individual process operations on our entire heat engine cycle, as shown on Fig. 3.1, we simply concern ourselves with the net heat engine work transfer between the system and the environment and the magnitude of the heat transfer of the engine with the hot and cold thermal reservoirs.

Figure 3.2: Heat Engine as a system in the environment showing work and energy transfers

This is shown in Fig. 3.2 where E is the entire heat engine of Fig. 3.1 say. The system-environment boundary is the circle around the E, T_h and T_c are the hot and cold thermal reservoir. A <u>thermal reservoir</u> is a very large thermal mass ensuring that whatever heat is added/taken from it at whatever rate the temperature of that thermal mass stays constant. Q_h and Q_c are the heat transfers to/from the hot and cold reservoirs respectively. W is the *net* work crossing the system boundary, e.g. $W_{out} - W_{in}$ from Fig. 3.1. A key simplification over the 1st Law is that the signs on Q and W are ignored here, the direction of heat and work flow is defined by arrows on the diagram. As we shall see in section 3.6 on page 46 it is the magnitude of the system heat transfers from and to the environment, Q_h and Q_c that define heat engine efficiency.

3.5   Reversibility for Cycles

In section 2.9.6 on page 37 single reversible and irreversible processes for systems were introduced. The system is a working fluid, a fixed mass that may change volume and do work, and/or heat may transfer to/from it to the environment. In comparison to section 2.9.6 on page 37 where we were concerned purely about the system undergoing the single process (ie how does the state of the fluid inside the system boundary change), now we are concerned about work or energy transfer between the system and the environment. Because we are discussing heat engine cycle performance, and from the 2nd Law in section 3.3, what state the environment is in is crucial to the engine performance. Therefore we have to include the state of the environment in the definition of process reversibility.

- A process is said to be <u>internally reversible</u> if no irreversibilities occur within the system during the process. Our typical quasi-equilibrium processes fit this description because as long as we can go from state 1 to state 2 along some process path, and go back down to state 1 again, then that process is <u>internally reversible</u>.
- An <u>externally reversible</u> process is one that causes no irreversible change outside the system boundaries. Heat transfers to thermal reservoirs are a good example since the temperature of the reservoir does not change. Note however this process would not be internally reversible if a temperature difference exists between the system and the environment.
- A <u>totally reversible</u> process involves no irreversiblities in either the system or the environment. This means the process must not involve a heat transfer with a temperature difference, it must be quasi-equilibrium and not involve any friction or other irreversible losses. A reversible adiabatic process is a good example.

3.6   Heat Engine Efficiency

Efficiency generally is "what you get" in terms of "what you pay for". Applied to heat engines: "what you get" is the net work output, W_{net}. "what you pay for" is the energy extracted from the high temperature reservoir. Note that the energy donated to the cold reservoir is wasted energy and is the source of our inefficiency. *All* Heat Engine Thermal Efficiencies are thus defined: $\eta_{th} = W_{net}/Q_h$. From the first Law for a cycle, eqn. 3.1, the net work can be defined in terms of the heat transfers, $W_{net} = Q_h - Q_c$ and therefore

$$\eta_{th} = \frac{W_{net}}{Q_h} = \frac{Q_h - Q_c}{Q_h} = 1 - \frac{Q_c}{Q_h}. \tag{3.2}$$

Clearly: the larger Q_h can be and the smaller Q_c can be the more efficient our heat engine will be. Practical inefficiencies, e.g. friction, decrease efficiency through reducing the net work.

3.6.1   Reversible Heat Engine Efficiency

One way to visualise a reversible engine is to ask yourself the question : could this engine run backwards around the cycle? If yes the engine is reversible. The efficiency of a reversible engine is dependent ONLY on the hot and cold reservoir temperatures. It does not depend on the type of processes, the working fluid or anything else. You can prove you can express η_{th} in terms of T_c and T_h (one needs an arrangement of 3 engines and some maths), but the important thing is

$$\eta_{th,rev} = 1 - \frac{T_c}{T_h}. \tag{3.3}$$

This instantly gives an idea of energy quality. If the hot and cold T are 1200K and 300K, $\eta_{th,rev} \sim 0.75$. Reduce the hot T to 600K and $\eta_{th,rev} \sim 0.50$. Clearly some energy sources are more efficiently used than others. The reversible heat engine efficiency is sometimes known as the Carnot Efficiency. As we shall see in section 3.8 on page 47, it is the efficiency that all practical engines aspire to.

3.7   Reversed Heat Engines and "Efficiency"

<u>Reversed</u> (not reversible) heat engines use work to move heat from a cold to a hot reservoir, and as sketched on a $p - V$ diagram, operate in an anti-clockwise fashion. Because of this the net work is negative and a classic example is a refridgerator, as shown in Fig. 3.3. We use electricity to drive a compressor to expand and cool a refridgerant to below T_c, which then extracts energy at the cold temperature. It is then compressed and heats up, and donates energy at a higher temperature (to the room the kitchen is in). Again efficiency is "what you get" in terms of "what you pay for". "What you pay for" is obvious: the electrical work to run the engine. What we get depends on the engine type. Also note "efficiency" here can be > 1, therefore the term is called "Coefficient of Performance". For a heat pumps: The useful energy is the hot heat flow (to a house): $COP = Q_h/W$. For refrigerators: The useful energy is the cold heat flow: $COP = Q_c/W$.

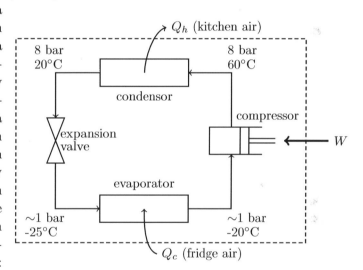

Figure 3.3: *A Reversed heat engine, taking in thermal energy from the cold reservoir, and using work to deliver it to the hot reservoir*

3.8   The Carnot Cycle: The Impossible Engine

The Carnot Engine has a frictionless piston-cylinder arrangement where the cylinder walls are perfectly insulated. The cylinder head has a removable perfect insulation. It may be replaced by a heat source or a heat sink. It is not, in any sense, a practical engine, rather a theoretical engine which is the most efficient any heat engine can possibly be. Understanding why a Carnot heat engine is efficient is to understand how practical heat engines described in section 3.9 have losses. Note that these losses in practical heat engines are not associated with practical issues like friction or leaks. These are losses associated with the fundamental inefficiencies in the way in which heat is transferred between the system and the environment. The Carnot Engine consists of 4 processes and is shown on $p - V$ and $T - s$ diagrams in Fig. 3.4. The four processes are:

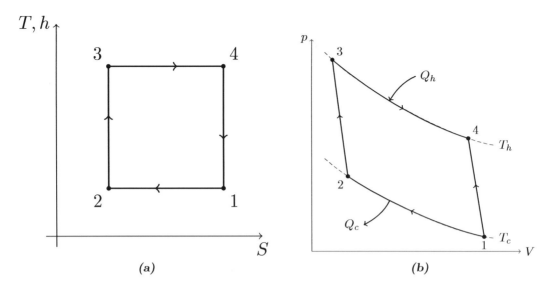

Figure 3.4: *The Carnot Cycle shown on (a) T-s and (b) p-V diagrams*

Process 1–2: Reversible isothermal compression by placing a heat sink in place on cylinder head. To ensure reversible compression process the temperature difference between the system and the environment must be zero and the process must be frictionless.

Process 2–3: Reversible adiabatic compression. To achieve this the heat sink is replaced with an insulated cylinder head, and again the process must be frictionless.

Process 3–4: Reversible isothermal expansion. Now the insulated cylinder heat must be replaced by a heat source. Same comments as process 1-2 apply with respect to reversibility.

Process 4–1: Reversible adiabatic expansion. Finally the heat source must be replaced again by an insulated cylinder head prior to the start of the process. Same comments as process 2-3 apply with respect to reversibility.

As is evident, the Carnot Engine is in no sense a practical engine. Although the adiabatic expansion/compression are reasonably well approximated in practice, the impossibility relates to the isothermal processes – notice heat transfer occurs <u>reversibly</u>. If $T_1 > T_2$ then heat can only flow from state 1 to state 2, and thus the transfer process is not reversible. The only way a reversible heat transfer can occur is if $T_1 \sim T_2$, and if this occurs no heat will flow. This means the heat transfer takes an extremely long time though an extremely large area, then the heat flow, from equation 2.12 on page 24 can occur over an extremely small temperature difference. The reason why the Carnot cycle has the maximum efficiency, as defined by equation 3.3 on page 47 is that all the heat transfer occurs at the heat sink/source temperature.

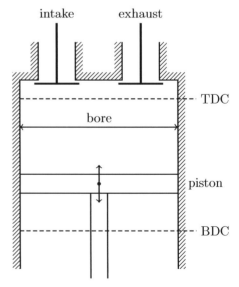

Figure 3.5: *Basic Features of a Practical Internal Combustion Engine*

All practical model systems as discussed in section 3.9 the heat addition and heat rejection steps occur over a temperature range, because the isothermal compression and expansion processes are, practically, impossible. This can be also understood graphically by examining the $T - s$ diagram for the Carnot cycle shown on Fig. 3.4b. As discussed in section 3.1 the First Law for a cycle tells us that the net work and the net heat transfer must equal. Therefore the maximum work possible between T_{max} and T_{min} is when the area enclosed is a maximum, i.e. a cuboid.

3.8.1   If a Carnot Engine is Totally Reversible, why is it not 100% Efficient?

Entropy changes occur for the two isothermal heat transfers and $s_{12} = q_{12}/T_1$ since q_{12} occurs at T_1. By reference to Figure 3.4b, $q_h = q_{34} = s_{34}T_3$ and likewise $q_c = -q_{12} = s_{12}T_1$, and from Figure 3.4a, $s_{34} = -s_{12}$. The key to understanding the Carnot efficiency is that q_h transfers entropy to the system at $T_h = T_3 = T_4$, and transfers the same amount of entropy to the environment at $T_c = T_1 = T_2$. So for the Carnot cycle there is *no* net entropy change of the environment, but there *is* a net heat transfer and that *net* heat transfer is 100% converted to net work. However from the point of view of the system, the conversion efficiency of the thermal energy *supplied* (q_h) is not 100% because the some thermal energy (q_c) is rejected. Think about this !

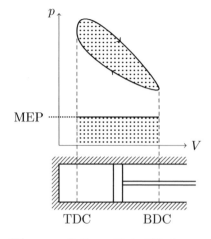

Figure 3.6: *Net Work expressed in terms of Mean Effective Pressure (MEP)*

Given these losses, and the innate limitations of (Carnot) heat engine efficiency, practical internal combustion engines are actually very efficient, and modern techniques have dramatically improved performance, fuel economy and emissions characteristics.

3.9   Practical Engines

There are a number of practical issues that limit the performance of a real engine compared to the Carnot engine.

- The hot reservoir temperature is limited by a number of factors: maximum combustion temperature possible, combustion temperature limitations due to combustion stability, emissions formation, mechanical failure of components at high temperatures.
- The cold reservoir temperature limit is defined by ambient conditions.
- Heat Transfer inefficiencies: heat transfer occurs over a finite temperature difference.
- Losses due to fluid mixing, turbulence (fluid viscous losses).
- Mechanical losses (friction, crevice flows).

Present day internal combustion engines have a vast scale and speed range, F1 engines typically run at 15000rpm, whilst marine Diesels have engine cylinders you can stand in. They all have the same basic characteristics as shown in Fig. 3.5. A piston moves a certain distance (stroke) between bottom and top dead centre (BDC and TDC). At TDC a clearance volume exists, V_{min} and the displacement volume, the volume swept by the piston and is defined as the cylinder volume at BDC – Clearance vol = V_{max}. The key geometrical design quantity of an internal combustion engine is the volume compression ratio $r_v = V_{max}/V_{min}$. Spark ignition engines use smaller compression ratios and lighter fuels and inject fuel various ways (through the port or directly into the cylinder) and a spark plug in the head ignites the mixture $\sim$ TDC. Compression ignition engines use larger compression ratios and inject heavier fuels directly into the cylinder under high pressure $\sim$ TDC to auto-ignite the combustion mixture.

As for any cycle the area enclosed on a $p - V$ diagram defines the net work done by that cycle. A common term to characterise engine performance is the mean effective pressure (P_{MEP}). It is a hypothetical pressure which, as shown by Fig. 3.6 if acted on the piston during the entire piston motion from BDC to TDC would produce the same net work obtained by the cycle. In other words the two shaded areas of Fig. 3.6 are equal and thus the MEP is defined $P_{MEP} = W_{net}/(V_{max} - V_{min})$.

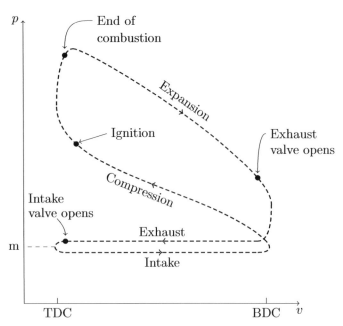

Figure 3.7: *A Practical 4-Stroke Spark Ignition Engine Cycle shown on a p-V diagram*

Figure 3.8 shows the cycle for a typical 4 stroke spark ignition internal combustion engine and Fig. 3.7 shows the cycle path on a $p - V$ diagram. Starting at the intake stroke exhaust valve closes, the intake opens, and a fresh cool air charge sucked into cylinder as the piston moves from TDC to BDC. This process consumes minimal power and the cylinder pressure is slightly under atmospheric pressure during this process.

For port injected spark ignition engines the liquid fuel is injected just upstream of the intake valve and mixes and evaporates during this process. For directly injected engines the fuel is either introduced to the cylinder during the intake stroke or the compression stroke depending on the mode of engine operation required. Once the piston reaches BDC the valves close and the compression stroke starts and the fuel-air vapour mixture is compressed. This is the stroke consuming the most power from the engine. Just prior to TDC the mixture is ignited and the combustion process is very fast, quickly increasing the cylinder pressure. Then the power stroke starts which extracts energy from the fluid. Near BDC the exhaust valve opens and the hot combustion gases are pushed out of the cylinder by the piston. This process, like the intake stroke, consumes minimal power but the cylinder pressure is slightly above atmospheric pressure.

Compression ignition engines have the same basic processes as spark ignition engines with a few key differences. One is that fuel is not injected during the intake stroke as in a petrol engine but injected near TDC. Another is there is no spark plug : the fuel auto-ignites when injected, and the combustion process is longer in duration since the fuel-air mixture is not pre-mixed. These differences provide a number of practical advantages for compression ignition engines.

Clearly real engine cycles are complex and the fluid mass undergoing the cycle changes every cycle. In theory this is an open thermofluid process, requiring a   control volume rather than a   system description, as discussed in section 2.3.5 on page 17. However with some not too drastic assumptions we can approximate engine cycles as closed systems, which enables us to define the cycle efficiency very simply, as described by section 3.6 on page 46.

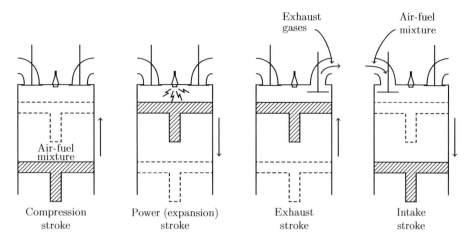

Figure 3.8: *The 4 strokes of a Practical Spark Ignition Engine*

3.10   Model Engine Cycles: Air Standard Assumptions

Using the typical 4 stroke spark ignition internal combustion engine cycle introduced above as a reference, we now define the assumptions used to create model cycles that use fixed fluid mass (a system) as a basis. These are:

- The working fluid is air which is contained in a closed cylinder and always behaves as an ideal gas.
- All the processes are internally reversible.
- The combustion process is replaced by a heat addition process from an external source.
- The exhaust process is replaced by a heat rejection process that restores the working fluid back to its original state.

These assumptions allow us to model real engines and gas turbines as systems undergoing a closed cycle of processes and have efficiencies defined by equation 3.2 on page 46. Sections 3.11 and 3.12 define the model cycles for spark and compression ignition internal combustion engines. Section 3.13 approximates a gas turbine as a closed system also, which is reasonable for a stationary gas turbine system. In section 4.11 on page 72 considers <u>open</u> thermofluid operations, we analyse the gas turbine operating as a jet engine using the <u>control volume</u> approach as a comparator.

3.11   The Otto Cycle: The Model Spark Ignition Engine

The Otto cycle approximates a real spark ignition engine using the following restrictions of the air standard assumptions given in section 3.10.

- The combustion process is replaced by a constant volume heat addition process, since the process process is fast due to the fuel-air premixing during the intake stroke.
- The exhaust and intake processes are replaced by a constant volume heat rejection process.
- The compression and expansion strokes are approximated by adiabatic processes.

Figure 3.9 shows the Otto cycle on $p - V$ and $T - s$ diagrams. The net work $= W_{12} + W_{34}$. $W_{23} = W_{41} = 0$. For an adiabatic process: $PV^n = \text{const} = K$ and therefore

$$
\begin{aligned}
W_{12} &= \int_1^2 p\,dV = K \int_1^2 V^{-n}\,dV = \frac{K}{1-n}\left(V_2^{1-n} - V_1^{1-n}\right) \\
&= \frac{1}{1-n}\left(p_2 V_2^n V_2^{1-n} - p_1 V_1^n V_1^{1-n}\right) = \frac{(p_2 V_2 - p_1 V_1)}{1-n}
\end{aligned}
\tag{3.4}
$$

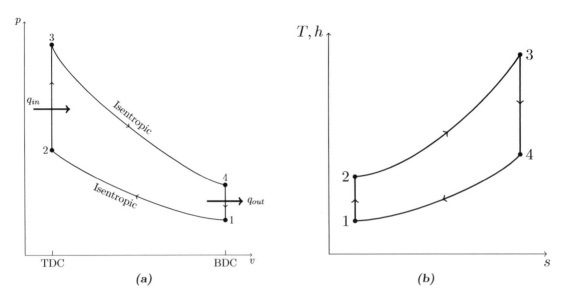

Figure 3.9: *The Otto Cycle depicted on (a) p-v and (b) T-s diagrams*

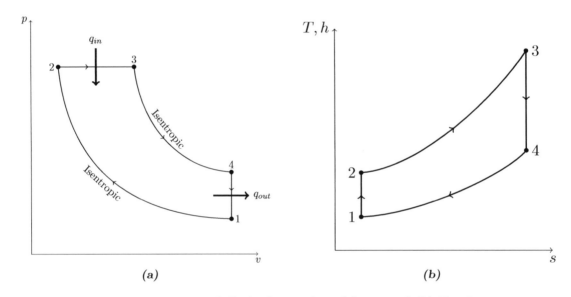

Figure 3.10: *The Diesel Cycle depicted on (a) p-v and (b) T-s diagrams*

Similarly $W_{34} = (p_4V_4 - p_3V_3)/(1 - n)$ and $W_{net} = (p_4V_4 + p_2V_2 - p_3V_3 - p_1V_1)/(1 - n)$. In terms of the heat transfers, the heat addition is $q_{23} = C_V(T_3 - T_2)$ and the heat rejection is $-q_{41} = C_v(T_4 - T_1)$. The efficiency of a heat engine is $\eta_{th,Otto} = w_{net}/q_{in} = 1 - q_{out}/q_{in}$. This gives $\eta_{th,Otto} = 1 - (T_4 - T_1)/(T_3 - T_2)$. This simplifies to

$$\eta_{th,Otto} = 1 - \frac{1}{r_v^{\gamma-1}} \tag{3.5}$$

where $r_v = v_1/v_2$ is the volume compression ratio. The Otto cycle efficiency in terms of compression ratio is shown in Fig. 3.11 for the $r_c = 1$ curve. It is a strong function of r_v for small r_v while at large r_v the improvements are not so great. Also, large r_v operation becomes difficult due pre-ignition of the fuel-air mixture, know as "knock" which will quickly lead to engine damage. Typically r_v is limited to ~ 10.

3.12 The Diesel Cycle: The Model Compression Ignition Engine

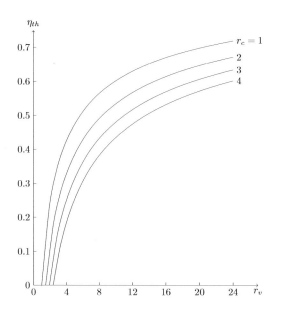

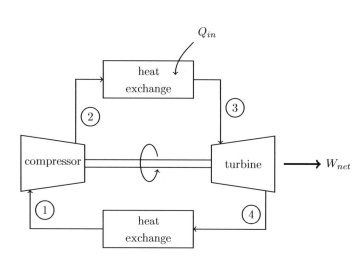

Figure 3.11: *Thermal Efficiency for Otto ($r_c = 1$) and Diesel ($r_c > 1$) Engine Cycles*

Figure 3.12: *A Gas Turbine modelled as a closed Brayton Cycle*

The Diesel cycle differs from the Otto cycle in only one respect, that the heat addition process is constant pressure rather than constant volume, and produces a slightly different path on a $p - V$ plot as shown in Fig. 3.10. Practically however, as noted in section 3.9 on page 49 and discussed more fully below this has important practical consequences. In terms of thermodynamic calculations, the heat addition process is now defined $q_{23} = C_p(T_3 - T_2)$ and, comparing to equation 3.4 for the Otto cycle efficiency a γ appears in the denominator, $\eta_{th,Diesel} = w_{net}/q_{in} = 1 - q_{out}/q_{in} = 1 - (T_4 - T_1)/\gamma(T_3 - T_2)$. An additional parameter, the "cutoff ratio", $r_c = v_3/v_2$, the relative volume change during the combustion process is used to define the Diesel cycle efficiency in terms of r_v. Note this is the same as the Otto cycle efficiency (equation 3.5) with the addition of the term in the square brackets, $\eta_{th,Diesel} = 1 - \frac{1}{r^{\gamma-1}} \left[\frac{r_c^\gamma - 1}{\gamma(r_c - 1)} \right]$. Since this term is always > 1, a Diesel engine is theoretically less efficient that a petrol/gasoline engine. The Diesel cycle efficiency in terms of compression ratio is shown in Fig. 3.11 for the $r_c > 1$ curves and the following comments can be made.

- As with the Otto Cycle it is a strong function of r_v for small r_v while at large r_v the improvements are not so great.
- Because in a compression ignition engine the fuel is injected after the compression process is complete, knock is impossible. This means that much larger compression ratios are possible in compression ignition engines. Typically r_v up to ~22.
- Because of the higher compression ratios they actually have a higher (practical) efficiency than spark ignition engines, which have a higher (theoretical) efficiency.
- Because the system is much less sensitive to early ignition, a wider and cheaper range of fuels can be used, from biodiesels, mineral oil to straight vegetable oil (the latter with preheating).
- Heavier fuels appropriate for compression ignition engines also have a larger energy content per unit mass.
- Because no ignition system is required for compression ignition engines, they are simpler and more robust, however the larger compression ratios used do require larger heavier engine cylinders.

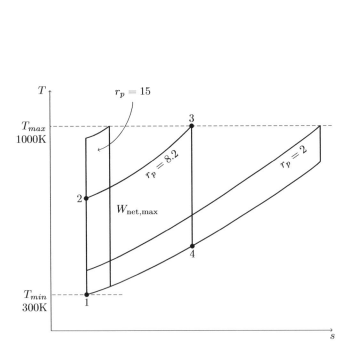
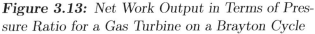

Figure 3.13: *Net Work Output in Terms of Pressure Ratio for a Gas Turbine on a Brayton Cycle*

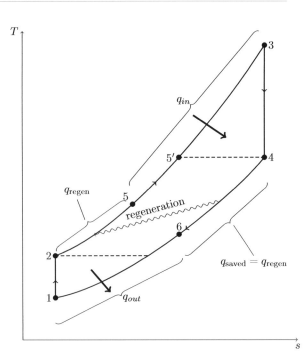

Figure 3.14: *Transfer of Thermal Energy from the turbine exhaust to the compressor intake of a Gas Turbine*

3.13 The Brayton Cycle: The Model Gas Turbine Engine

As with internal combustion engines gas turbines usually operate on a open cycle where the turbine drives the compressor and provides the remainder of shaft work to the environment, for instance an electrical generator. Fresh air is drawn in, combustion fuel mixed, ignited and the exhaust gases discharged – in the case of a jet engine through a nozzle. As shown in Fig. 3.12 a gas turbine can be approximated as a closed cycle, called the Brayton (or Joule) cycle using these restrictions and air-standard assumptions noted in section 3.10 on page 51.

- The combustion process is replaced by a constant pressure heat addition.
- The exhaust is replaced by a constant pressure heat rejection process.
- The compression and expansion processes are again assumed to be adiabatic.

In terms of heat transfer processes note the Otto cycle (section 3.11) has two constant volume processes, the Diesel cycle (section 3.12) has one constant volume and one constant pressure process and the Brayton cycle has two constant pressure processes.

The Brayton Cycle thermal efficiency is the same as that of the Otto cycle (equation 3.4) because in the heat addition and rejection processes are the same within each cycle. However the efficiency is usually defined in terms of a pressure (compression) ratio where $r_p = p_2/p_1$ and becomes $\eta_{th,Brayton} = 1 - 1/r_p^{(\gamma-1)/\gamma}$ rather volume (compression) ratio as in the Otto cycle.

In terms of thermal efficiency, everything that has been said about Otto cycle on page 51 applies to Brayton cycles, since the efficiency definition is the exactly same. In terms of range, for stationary power systems the pressure ratio is typically 5-20. The performance limitation of a gas turbine is T_3, the hot combustion gases from the burner going into the turbine, due to the mechanical integrity of the turbine blades. As noted by Fig. 3.13, for a fixed turbine inlet temperature (T_3) the work obtained from the turbine increases at small r_p, to an optimum, before finally decreasing again at

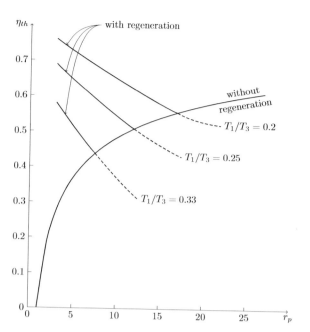

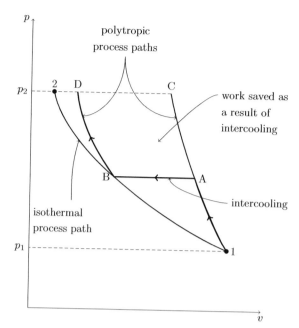

Figure 3.15: *Effect of Regeneration on the Thermal Efficiency of the Brayton Cycle*

Figure 3.16: *Effect of Staged Compression on the Work Required to Compress a gas from state 1 to state 2*

high r_p. Therefore a decision is required for a gas turbine at the design stage. High pressure ratios, as shown by Fig. 3.11 give good thermal efficiency but not the best specific work, as shown on Fig. 3.13.

3.13.1 The Brayton Cycle with Regeneration

A considerable amount of the waste heat from the turbine exit stage (T_4), can be recovered by using it to preheat the fluid entering the combustion chamber as shown in Fig. 3.14 and this process is known as regeneration. If the temperature leaving the turbine, $T_4 > T_2$, the temperature leaving the compressor (T_2) can be increased. It can be useful to preheat the inlet feed to the combustion chamber (T_5). If the regenerator was 100% efficient (in other words it was infinite in size and had no thermal resistance), $T_5 = T_4$, however a more practical and economically viable efficiency is 70%. With this value, as noted in the $T - S$ diagram, q_{out} is reduced by q_{saved} and q_{in} is reduced by q_{regen}. Regeneration works best when the pressure ratio is low as shown by Fig. 3.15.

3.13.2 The Brayton Cycle with Reheating and Intercooling

The efficiency of the entire cycle can also be improved using staged compression and expansion, with intercooling or reheating respectively. Intercooling is constant pressure heat removal. This increases the gas density and the compression efficiency. Reheating is constant pressure heat addition. How the process works is explained with reference to Fig. 3.16, for intercooling, and the aim of the compression process is to get from state 1 to state 2. Using single stage compression we compress the gas from state 1 to state C and then cool the gas to get to point 2.

In the two stage process we go from state 1 to state A in the first stage, cool the gas at constant pressure to state B and then take the gas pressure up to state D using the second stage compression, and a final small heat rejection process to get to state 2. The work done by the single stage compressor is the area under 1C2, whilst the area under the two stage process is 1ABD2, clearly less. Practically, the work saved must be balanced against in a more complex multi-stage system, with more friction and other losses. Two or three stages are typical.

3.14   Entropy, 2nd Law and Useful Energy

The difference in thermal efficiency between a Carnot engine (the most efficient heat engine possible) and those of practical engines is due to the loss of available energy due to non-isothermal heat transfer, and is now explained. Consider a pointless but <u>totally</u> reversible heat engine which has 2 processes that operate between states 1 and 2 :

- Process $1 \to 2$: Isothermal Compression.

- Process $2 \to 1$: Isothermal Expansion.

This is a pointless engine because the net work and the net heat transfer over the cycle is zero, but serves to introduce the basic concepts before going onto a pointless internally reversible engine. For process $1 \to 2$ we know from the First Law, for the system, $w_{12} = q_{12}$ and $w_{12} = \int pdv = p_1 v_1 ln(v_2/v_1)$. We know from the Second Law, for the system, that $s_{12} = q_{12}/T_1 = Rln(v_2/v_1)$ and the entropy change will be negative. Because the process is isothermal and reversible, the temperature of the system and the environment is the same. Therefore $s_{12}|_{SYS}= -s_{12}|_{ENV}$. Also, because $2 \to 1$ is the exact reverse of process $1 \to 2$ then the same applies for the thermal, work transfers. Crucially, we can confirm in this case that neither the entropy of the system nor the environment has changed over the cycle. Of course the entropy of the *system* cannot change over the *cycle*, so what we have shown here is neither does the entropy of the environment, eg $\oint dS \,|_{SYS}+ \oint dS|_{ENV}= 0$ and thus this engine is *totally* reversible.

Now consider an equally pointless but <u>internally</u> reversible heat engine has two different processes :

- Process $1 \to 2$: Constant Volume Heating.

- Process $2 \to 1$: Constant Volume Cooling.

As before, for process $1 \to 2$ from the First Law, for the system, $q_{12} = C_v(T_2 - T_1)$ and from the Second Law, for the system, $s_{12}|_{SYS}= Rln(T_2/T_1)$. Both the heat transfer and the entropy change, with respect to the system is positive. Now let us consider the environment, and let us say the environment is a thermal reservoir at T_2, since the system was heated to T_2. So the entropy change of the environment is simply $s_{12}|_{ENV}= -q_{12}/T_2 = -C_v(1 - [T_1/T_2])$, and of course this is negative because with respect to the environment, the heat transfer was *from* it. Now let us say $T_2 = 2T_1$, so $s_{12}|_{SYS}\sim +0.693C_v$ and $s_{12}|_{ENV}= -0.5C_v$. Therefore for process $1 \to 2$ of our cycle, the entropy of the system has increased and environment has decreased. Notice though, there is a net entropy gain for process $1 \to 2$ adding the two contributions.

We still have to get our system back down to T_1 again in process $2 \to 1$ and to do that we have to use another thermal reservoir in the environment, with a temperature of T_1. Now, for process $2 \to 1$ from the First Law, for the system, $q_{21} = C_v(T_1 - T_2)$ and from the Second Law, for the system, $s_{21}|_{SYS}= Rln(T_1/T_2)$. Both the heat transfer and the entropy change, with respect to the system are negative and are of the same magnitude as in process $1 \to 2$. By the same argument as previously for the environment, for process $1 \to 2$, $s_{21}|_{ENV}= -q_{21}/T_1 = -C_v(1 - [T_2/T_1])$, and for $T_2 = 2T_1$, $s_{21}|_{ENV}= C_v$. Again, overall, for process $2 \to 1$, there is a net gain in entropy. By definition, $\oint dS|_{SYS}= 0$ because the process is internally reversible, but here $\oint dS|_{ENV}> 0$. Notice the following.

- During process $1 \rightarrow 2$ the total entropy of the system + the environment increased.

- During process $2 \rightarrow 1$ the total entropy of the system + the environment increased.

- Over the cycle the entropy of the environment increased.

These two examples are pointless engines because they generate no net work, but Carnot and Otto engines are precisely these heat transfer processes with additional (and isentropic) adiabatic compression/expansion stages. And, let us not forget (1) it is the heat transfer processes that cause entropy change since $\delta s = \int \delta q / T$ and (2) Carnot and Otto engines generate net work. This is why an Otto cycle has a lower thermal efficiency than a Carnot cycle.

Practical engines increase the entropy of the environment, due to the externally irreversible constant volume or pressure heat transfer processes. If we take some reference temperature of the environment as T_{env} say, then the additional heat transfer to the environment due to irreversibility could be expressed as $T_{env} \oint dS|_{env}$. This represents an addition to the q_c term in the thermal efficiency (equation 3.2 on page 46) such that $q_c = q_{c,rev} + T_{env} \oint dS|_{env}$, which is another way to link entropy generation to efficiency loss in heat engines.

3.15   Key Points

We have now considered energy conservation for a fixed mass of a compressible ideal gas and examined how efficiency can be linked to minimal entropy generation. In summary,

- A Cycle of Processes is used by a heat engine to convert thermal energy into work.
- In doing so the heat engine must reject thermal energy to the environment, causing a fundamental inefficiency.
- The Carnot Engine is the most efficient possible, practical engines have lower efficiencies because the heat exchange occurs over a temperature difference.
- Otto Cycles are more efficient than Diesel cycles but the latter are more efficient practically.
- Practical methods such as staged compression/expansion and regeneration can improve theoretical efficiency.

4 | Rate of Mass, Momentum and Energy Conservation for a Fixed Volume

We now consider the dynamics of a thermofluid system where bulk flow *is* present, and as discussed in section 2.3.1 on page 15 this requires conservation over a fixed volume, rather than the fixed mass (system) approach used in chapter 3. Here we describe the motion of fluid elements, and how the conservation of mass, momentum and energy may be expressed for a control volume. We make one key assumption in this chapter, that the fluid is inviscid and thus a direct balance between force and acceleration (rate of change of momentum), as defined by Newton's Second Law. It is a usually a good approximation to thermofluid processes away from walls. This force can come from gravity or from a change in pressure in the direction the fluid is flowing. This is a partial treatment (without the viscous forces of chapter 5) of the full conservation equations, given in chapter 8 on page 111.

4.1 Methods to Visualise Fluid Motion

Fluid Mechanics is an old subject, and scientists and engineers have always undertaken experiments to study fluid motion in order to develop physical understanding and show that theoretical equations are true. Three methods of visualizing fluid motion are commonly used, Pathlines, Streaklines and Streamlines. The first two have distinct similarities, but the latter has a completely different basis. They all however create the same visualisation of the fluid motion when the fluid flow is steady.

4.1.1 Pathlines and Streaklines

A pathline is a 'path' traced out by a *single* fluid particle released at a specific point in space $\vec{x}_0$ and at a specific point in time t_0. It is best thought of as a massless particle, and moves with whatever velocity the fluid has at the particle location at that time. Therefore the equation of motion for the pathline is $D\vec{x}/Dt = \vec{u}(\vec{x}, t)$ where $\vec{u}$ is the velocity of the fluid at the particle location. A pathline is the complete position history of a single massless particle.

A streakline on the other hand is the time history of a series of particles all starting from the same position but at *different* times. The best way to think of a streakline is the trace of a line of dye injected into a moving fluid by a stationary injector. Pathlines and streaklines are identical in steady flows if the particle creating the pathline starts from the dye injection location. However, in unsteady flows, the resulting 'trace' can be completely different, as is explained in the following example.

Figure 4.1a shows a uniform flow from right to left at $t = 0$, represented by the horizontal stream*lines*. At this time one particle is placed in the fluid at the position shown by the solid circle, and also a dye injector is placed above it, defined by the solid cube in the figure. At some time Δt later, as shown in Figure 4.1b the particle has moved to the new position of the solid circle, but remember the old positions are part of the pathline trace, so we show these as open circles. In the same time interval the dye streak emerged from the injector, following the flow in the same direction, and this is shown also. Now we assume that at $t = \Delta t$, just after the 'photo' in Figure 4.1b was taken, the flow direction instantly changes to that shown in Fig. 4.1c and stays that way until $t = 2\Delta t$. From $t = \Delta t$ the flow moves vertically, and so does the particle, and it ends up in the final position shown in

(a)   (b)   (c)

Figure 4.1: *The difference between Pathlines and Streaklines in a simple unsteady flow where in (a) and (b) the flow is horizontal and in (c) the flow is vertical*

Fig. 4.1c, shown by the solid circle. Again the position history of this particle is recorded with some more open circles. Considering the streakline, the new dye injected between $t = \Delta t$ and $t = 2\Delta t$ will move vertically, because that is the direction of flow. *However* the old dye, injected between $t = 0$ and $t = \Delta t$ will *also* move vertically, because that is the direction of flow in the time interval $t = \Delta t$ and $t = 2\Delta t$.

4.1.2   Streamlines and Streamtubes

Streamlines and the closely related streamtubes are completely different from pathlines and streaklines. Imagine you have taken a photo of a flow, and you know what the velocity is everywhere in that photo. What you have is the velocity vector *field* at an instant in time. Now imagine you define some position in your photo, and you then draw two lines emanating in opposite directions from your point and always tangential to the fluid velocity. This is a point on a streamline, and, obviously, it only has any meaning for a steady flow.

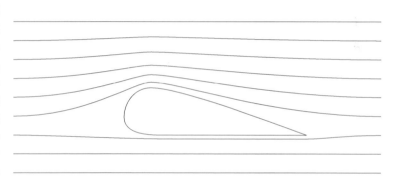

Figure 4.2: *A 2D streamline/tube plot around a streamlined object, highlighting fast and slow flow speed regions*

The thin lines in Figure 4.1a are all streamlines. It is also clear they are completely different from pathlines and streaklines, which are different types of time histories of particle motion. In 2D $x - y$ coordinates, streamlines are defined by,

$$\frac{dx}{dy} = \frac{u}{v}. \tag{4.1}$$

Since a streamline at a point S along it is tangential to the flow at that point they cannot cross each other. If they were to cross this would indicate two different velocities at the same point in space, which is impossible.

A stream tube in 3D is best imagined as a flexible tube whose length wise surface is defined by a collection of streamlines. Since the flow is tangential to this surface, mass flow down the tube at any point along it may be defined $\dot{m} = \rho A u_S$ where u_S is the uniform flow speed along the streamtube axis and A is the cross-sectional area of the tube at that point. For constant density fluids therefore the stream tube cross-sectional area is inversely proportional to the velocity magnitude in the direction of the tube. This is extremely useful when examining a streamline plot of a flow field, for instance in the 2D flow shown in Fig. 4.2 since we can visually see where the flow is accelerating, where the streamlines come together. As shown in the figure the flow is slow before and aft of the object but fast above and below it at the objects widest point.

4.2   Momentum Conservation along a Streamtube : The Euler Equation

Here we are going to apply Newton's 2nd Law (equation 2.8 on page 15) to an volume element of a streamtube, for instance those shown on Fig. 4.2. We assume two forces act, the pressure change along the streamtube and acceleration due to gravity.

Note we are assuming no viscous forces are present since the fluid is inviscid as noted in the introduction of this chapter on page 58. Acceleration is a force (force per unit mass) so in words our force balance on a fluid element along the stream tube is: (mass x acceleration) = (pressure change x area) + (mass x gravity). Figure 4.3 shows a sketch of the streamtube element and the relevant variables.

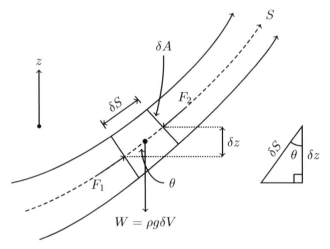

The 1-D Coordinate direction defined along the streamline is S and fluid acceleration may be converted as follows. $du_s/dt = du_s/dS \cdot dS/dt = u_s du_s/dS$. The volume element has length δS and area δA hence volume $\delta V = \delta S \, \delta A$. The weight of the fluid (N) is $mg = \rho \, \delta V \, g$ acting in the z-direction, and $\rho \, \delta V \, g \cos\theta$ acting in the

Figure 4.3: *A force balance on a small element of fluid in a streamtube*

S-direction. The pressure force at $S = 0$ is $F_1 = p \, \delta A$, while the pressure force at $S = \delta S$ is expanded as a Taylor series and is $F_2 = (p + dp/dS \delta S) \, \delta A$. Because the flow is steady (it is a streamtube!) the forces must be in equilibrium and can be equated : $\rho \, \delta S \, \delta A \, u_s du_s/dS = p \, \delta A - (p + dp/dS \, \delta S) \, \delta A - \rho \, \delta V \, g \cos\theta$. If we divide this equation by $\delta S \, \delta A$, define $\cos\theta = \delta z/\delta S$, and take the limit as $\delta S \to 0$, $\delta z \to 0$ (by reference to section 2.2.4 on page 11), we obtain a differential force balance for any point S somewhere along the streamtube.

$$\rho u_s \frac{du_s}{dS} + \frac{dp}{dS} + \rho g \frac{dz}{dS} = 0 \tag{4.2}$$

This is a form of the Euler equation, and it defines the force balance in an inviscid fluid, here along a streamline. This form tells us that total force change is zero along a stream tube, but it can transfer between forms. Note how the pressure gradient is an important force in fluid motion. Think again : the pressure in the fluid does not affect the fluid motion, only the change of pressure from one point in the fluid to another. This is subject to another assumption of our streamlines : incompressibility. We are accustomed to thinking of pressure as a thermodynamic variable, linking pressure and temperature by equation 2.17 on page 32. However, if the fluid is incompressible, this link is lost, and a different understanding is required. It is also important to recall the assumptions, steady flow of an incompressible inviscid fluid. So this means that if the fluid is inviscid and incompressible and the streamtube is normal to the gravity direction, then only a pressure gradient will make the fluid move.

4.3   Mechanical Energy Conservation along a Streamtube: The Bernoulli Equation

If we integrate equation 4.2 with respect to S we obtain an energy balance (since Energy = Force × distance) along the streamline. Since we have not included any thermal energy components and assumed our streamtube is incompressible this is a pure mechanical energy balance. The spatially integrated form of equation 4.2 is known as the Bernoulli equation and is defined

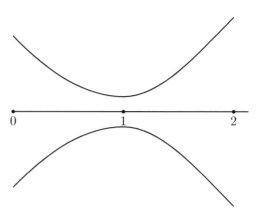

$$\frac{\rho u_s^2}{2} + p + \rho g z = const \qquad (4.3)$$

This tells us that the total energy is constant along a stream tube, but can change from kinetic, pressure and potential energy forms. This is the form most used in gas flows. In liquid flows, we divide by ρg and talk in terms of energy 'head'

Figure 4.4: *Caviatation : An application of the Bernoulli Equation*

which has dimensions of [L], $\frac{u_s^2}{2g} + \frac{p}{\rho g} + z = H$. Remember, again, we have made several implicit assumptions by using a streamline as a basis:

- The flow must be steady – otherwise streamlines have no meaning.
- There are no viscous forces.
- The fluid is incompressible.

These assumptions are key to understanding where NOT to apply the Bernoulli Equation. The first two are implicit in the meaning of a streamline and the forces used in the balance (equation 4.2) respectively. The third assumption can be relaxed (section 10.2 on page 127) but ensures the energy balance now includes thermal effects.

4.3.1   Cavitation: An application of the Bernoulli Equation

If we consider a flow through a narrowing channel, for instance as shown in Fig. 4.4, and considering the Bernoulli equation, equation 4.3, we can write $\rho u_1^2/2 + p_1 = \rho u_2^2/2 + p_2$. As the flow velocity increases, the pressure decreases, and for a given stagnation pressure (when $u = 0$) a maximum velocity exists. Boiling may be approached by either increasing the temperature (increasing the vapour pressure in the liquid), decreasing the pressure above it, or both (boiling a kettle on top of a high mountain). Cavitation due to flow is characterised by a Cavitation Number, $Ca = (p_{local} - p_v)/(1/2\rho u^2)$ and arises where the local fluid pressure reduces to the vapour pressure of the liquid.

4.4   A Note about Pressure

The variable 'pressure' has a number of different meanings in thermofluids and it is important to understand they are *ALL* valid interpretations in the *CORRECT* context.

4.4.1   Defining Mass and Density

Here we have to be mindful of the fluid compressibility assumption. In the limit of a low pressure compressible gas, then the gas law (equation 2.17 on page 32) shows us that at a given temperature the pressure defines the gas density, or mass per unit volume. So in this case pressure relates directly to mass per unit volume. In the other limit, an incompressible liquid, then the density is constant and the value of density is independent of pressure.

4.4.2   Defining Force

Here we have to be mindful of scalar and vector properties. Pressure has dimesions of force per unit area, or stress. However pressure is a scalar variable defining magnitude only, whilst force is a vector. Therefore we should consider pressure a force magnitude per unit area. It is the direction of the area normal that gives us the vector force, such that $\vec{F} = p\vec{A} = pA\vec{n}$ where $\vec{n}$ is the unit normal of the area. In stationary fluids the hydrostatic pressure in the liquid creates a force on the wall area of the container.

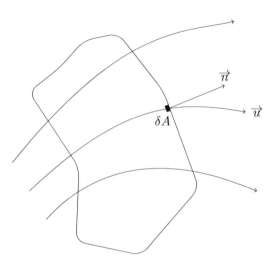

Figure 4.5: *A 2D control volume with a set of streamlines passing through it*

For inviscid incompressible fluids in motion, the force balance along a streamline is defined by Euler's Equation (section 4.2 on page 60). Note that the value of the pressure itself is irrelevant here, it is only the change in pressure along the streamline that creates a force *density*, since dp/dx has units Nm^{-3}. This means for example if one has liquid flowing out of a nozzle due to a pressure difference of 2 bar across it, the absolute value of the pressure in the fluid is irrelevant. The flow due to a pressure change from 3 to 1 bar and 30 to 28 bar would be exactly the same.

4.4.3   Defining Energy

Pressure is also an energy measure by virtue of $E = Fx$ and is most understood by writing the Bernoulli Equation (equation 4.3 on page 61) from a stagnation point, $p_0 = p_1 + (1/2)\rho u_1^2$. Here p_0 is the stagnation pressure and is the total (pressure) energy present in the fluid when it is at rest. When the liquid starts moving it picks up kinetic energy and the local (dynamic) pressure energy reduces. Note this is only true for incompressible fluids, for compressible fluids we must use stagnation enthalpy instead in the simplified form of the steady flow energy equation (section 4.5 on page 64).

4.5   Eulerian Conservation Equations

As discussed in section 2.3.1 on page 15 our (Lagrangian) fundamental conservation laws for mass, momentum and energy are defined in terms of a fixed mass. We as engineers however find conservation laws over (Eulerian) fixed volumes much more useful. The Eulerian basis is not *fundamental* in the sense they are not conserving fluid properties over the same collection of molecules (the fixed, system, mass). Converting conservation laws from a Lagrangian to an Eulerian basis is easy, we simply use the Reynolds Transport Theorem (section 2.3.6 on page 17) and indeed this is exactly what we do in chapter 8 on page 111. Here however, we discuss the process through examples.

The general conservation of mass, in 3D, assuming zero mass creation, for instance from chemical reactions is "Rate of accumulation of mass in the volume + net rate of mass leaving the volume = 0".

If the volume has a volume δV, then the mass in the volume at a point in time is $\int \rho \, \delta V$, and the rate of change of this is $\partial/\partial t(\int \rho \, \delta V)$. So the fundamental equation for the conservation of mass is:

$$\frac{\partial}{\partial t}\left(\int \rho \, \delta V\right) + \int \rho \left(\vec{u} \cdot \vec{n}\right) \delta A = 0. \tag{4.4}$$

For incompressible fluids, we can also write a volume conservation equation $\partial/\partial t(\int \delta V) + \int (\vec{u} \cdot \vec{n}) \, \delta A = 0$. Very often the flow velocity across inlets and outlets can be considered uniform and stated directly in terms of the normal velocity component. If in addition the flow can be considered steady then the mass conservation equation simplifies considerably, $\sum\limits_{inlets} \rho u_n A = \sum\limits_{outlets} \rho u_n A$ and similarly for the volume conservation equation.

Note equation 4.4 is exactly the same result as taking the fundamental equation for mass conservation (equation 2.7 on page 15) and applying the Reynolds Transport theorem (section 2.3.6 on page 17) for $(Z, z) \equiv (m_{sys}, \rho)$. The key point, made earlier in section 2.3.6 is that Z can be any variable, and here $Z \equiv m_{sys}, \vec{M}, E_T$, and we transform our fundamental mass, momentum and energy equations (2.7 to 2.9) from a Lagrangian (system) to an Eulerian (control volume) basis.

4.6 Conservation of Mass

Imagine we have some arbitrary steady 2D flow field, defined by a set of streamlines. Now suppose we draw an imaginary area somewhere in this flow field. Now suppose we want to ensure mass is conserved within this area. We do this as shown Fig. 4.5, by defining the mass flow through a tiny elemental area δA. At this tiny elemental area δA, the velocity vector $\vec{u}$ passes through it at a different angle to the unit normal of the area, $\vec{n}$, a vector of unit magnitude.

This is shown in Fig. 4.5. Conceptually, the mass flow through the area element is due only to the normal component of the velocity so what we want to do is find out how much of $\vec{u}$ is going in the $\vec{n}$ direction. From simple trigonometry the normal component is $|\vec{u}|$, or, using the dot product (equation 2.3 on page 10) $\vec{u} \cdot \vec{n} = |\vec{u}| \cos\theta$. So the mass flow through the small element is $\delta \dot{m} = \rho \left(\vec{u} \cdot \vec{n}\right) \delta A$. Note that when the angle $\theta < \pi$ radians, $\delta\dot{m} > 0$, and when $\theta > \pi$, $\delta\dot{m} < 0$, so the sign of the mass flow corresponding to mass inflow and outflow are automatically taken care of. Also note that when $\theta = \pi$, $\delta\dot{m} = 0$, here the flow velocity is tangential to the surface and the mass flow into/out the surface is zero. Since our flow is steady, the sum of the elemental mass flows into the control volume is the sum of all the elemental mass flows out of the control volume. For steady flows the net mass flow must be zero, $\int \rho \left(\vec{u} \cdot \vec{n}\right) \delta A = 0$.

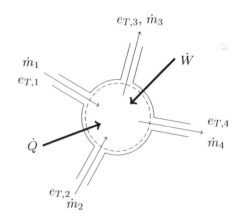

Figure 4.6: *Conservation of Energy for an Inviscid Fluid*

4.7 The Steady Flow Energy Equation (SFEE)

In section 2.7.7 on page 33 we defined the First Law of Thermodynamics fundamentally for a collection of molecules and in section 3.1 on page 44 we used it to analyse the energy changes for a fixed mass of fluid. What we did was define a system (a fixed mass in a Lagrangian Coordinate system) and conserve energy in that system in terms of energy and work transfers between the system and the environment as defined by equation 2.9 on page 15. In words our energy conservation rate equation is,

$$
\begin{bmatrix} net\ rate\ of\ thermal \\ energy\ transfer\ across \\ system\ boundary\ (J/s) \end{bmatrix} - \begin{bmatrix} net\ rate\ of\ work\ the \\ system\ does\ on\ its \\ environment\ (J/s) \end{bmatrix} = \begin{bmatrix} net\ rate\ of\ change \\ of\ energy\ of\ the \\ system\ \ mass\ (J/s) \end{bmatrix}
$$

Remember the fundamental conservation equations must be based on a Lagrangian fixed mass, equation 2.9 on page 15. The quantities in this equation are all <u>extensive</u> variables, for a <u>system</u> of fixed mass. The sign convention for thermal energy transfer is positive into the system and for work energy transfer is positive out of the system, as previously (section 2.7.8 on page 33). Here D/Dt is the Lagrangian (Material) Derivative (equation 2.6 on page 14), and here we are only interested in the steady part of this. E_T (J) is the total energy, in specific form, $e_T = e_u + e_k + e_p + e_g$. The specific kinetic, $e_k = u^2/2$, pressure $e_p = p/\rho$ and potential, $e_g = gz$, energies are part of the mechanical energy description we encountered in the Bernoulli equation (equation 4.3). The extra energy component present here is the specific internal energy $e_u = C_v T$. Note however, and this is very important as we shall see in section 5.5.1, the Bernoulli equation and the SFEE are defined using a different basis. The Bernoulli equation an energy balance along a streamline, and inherits the restrictions inherent in a streamline assumption, whereas the SFEE makes no such assumptions.

We want to conserve energy in a fixed volume through which mass and energy pass through (Eulerian description) rather than a fixed mass that moves with the flow (Lagrangian description). We do this because we are normally interested in working out energy budgets for engineering equipment, for instance nozzles, turbines, boilers, heat exchangers etc. To do this we operate on equation 2.9 first we define the RHS as energy and work transfer to the control volume boundary. Next we apply the steady part of the Reynolds Transport Theorem (page 17) to the LHS, e.g. $D(E_T)/Dt = \int \rho e_T (\vec{u} \cdot \vec{n}) \delta A$. This gives $\dot{Q} - \dot{W} = \int \rho e_T (\vec{u} \cdot \vec{n}) \delta A$. This equation is completely general for control volumes, fixed in space and steady flow. Substituting the energy components for e_T gives

$$
\dot{Q} - \dot{W} = \int \left(C_v T + u^2/2 + p/\rho + gz \right) \rho \vec{u} \cdot \vec{n} \delta A \tag{4.5}
$$

It is important to note *all* these energies on the RHS *flow* into and out of the control volume by convection of the fluid. This seems obvious for thermal and kinetic energy, but less obvious for potential and pressure energy. For potential energy the best way to think about it is the net change in height between an inlet and an outlet of a control volume, there is a net potential energy change to the fluid in the control volume. Often internal and pressure energies are added to define the specific enthalpy (equation 2.18 on page 32). Because C_p is usually known, this simplifies energy calculations of flowing compressible ideal gases considerably. Equation 4.5 defines a control volume based energy rate equation appro-

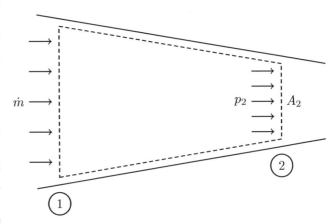

Figure 4.7: *Flow through a converging section*

priate for open processes and importantly contains no other assumptions other than the condition of steady flow. If we assume the flow is inviscid and the flow velocity is uniform across all outlets and inlets then the discrete form would be

$$
\dot{Q} - \dot{W} = \sum_{out} \dot{m} e_T - \sum_{in} \dot{m} e_T \tag{4.6}
$$

since $\dot{m} = \int_A \rho \vec{u} \cdot \delta \vec{A}$ and would represent the conservation of the energy rate of change as shown in Fig. 4.6, where the dashed line represents the control volume boundary. The $\dot{Q} - \dot{W}$ terms on the LHS of equation 4.6 are strictly heat and work transfers across the control volume boundary, in exactly the same manner as the system description in chapter 3 on page 44. Therefore $\dot{Q}$ is thermal energy transferred to the fluid due to external heat, cooling or combustion processes for instance. Similarly $\dot{W}$ is the work done on or by the fluid, for instance increasing the fluid pressure due to a pump or compressor, or energy extracted from the fluid, for instance by a turbine. Note further that $\dot{Q}$ is *strictly* thermal energy transfer, and for cases such as electrical heating, the definition of whether this is $\dot{Q}$ or $-\dot{W}$ depends on where you place the control volume boundary.

4.7.1   Pressure Energy Convected by the Flow

Pressure energy convection deserves some thought. Here we examine the flow through a converging section as shown by Fig. 4.7. We use the idea that work = energy = force × distance. At point 2, the outlet, the fluid inside the CV pushes on the fluid outside the CV with a force $F_2 = p_2 A_2$. During a time interval δt the fluid displacement is $\delta x_2 = u_2 \delta t$ so the amount of flow work is $\delta e_{p,2} = F_2 \delta x_2 = p_2 A_2 u_2 \delta t$ and the rate of flow work is therefore $\delta e_{p,2}/\delta t = p_2 A_2 u_2 = (p_2/\rho)\rho A_2 u_2 = (p_2/\rho)\dot{m}$. This energy transfer is positive because the work is being done by the control volume fluid on the environment. At 1, the inlet, the term is negative because the fluid outside of the CV is doing work on the fluid inside the control volume. This is completely compatible with our independently derived steady flow energy equation, equation 4.5. Note that several texts treat this 'pressure energy convection' as a form of external work, and add it to the LHS of 4.5 or even to the $\dot{W}$ term itself. This creates ambiguities since the LHS of 4.5 is strictly thermal and work transfers to the fluid in the control volume *not* due to convection and the RHS is energy transfers to the fluid in the control volume *by* the flow of the fluid.

4.7.2   Neglecting Negligible Energy Components

Often we can neglect some of the energy components, sometimes it is obvious, but we can be systematic. If we for instance assume a compressible fluid as an example, and we use enthalpy then the total specific energy in a flow stream is $C_p T + u^2/2 + gz$. As in section 2.6.5.5 on page 29 we split each variable into a dimensional and a non-dimensional part, for instance $C_p = C_{po} C_p^*$, where the non-dimensional part is of order unity. Other terms are $u = u_o u^*$, $T = T_o T^*$, $g = g_o g^*$, $z = z_o z^*$. Note C_p, g are constant so $C_p^* = g^* = 1$. We also make use of $C_p/C_v = \gamma$, $C_p - C_v = R$, $C_p = \gamma R/(\gamma - 1)$ and so the non-dimensional form can be written, $T^*/((\gamma_o - 1)Ma^2) + u^{*2}/2 + z^*/Fr^2$ where $Ma = u_o/(\gamma_o R_o T_o)^{1/2}$ and $Fr = u_o/\sqrt{g_o z_o}$.

4.7.3   Energy Conservation in Common Process Operations using the SFEE

Common process operations usually have considerable simplification of the SFEE, ie equation 4.6, and to be systematic you can apply the principle outlined in section 4.7.2 to check. Table 4.1 shows common process operations and the simplified SFEE in each case.

- For the <u>burner</u> the control volume is drawn to enclose only the fluid which is being heat by a burner. There are 2 ports, $\dot{W} = 0$, Potential and Kinetic energy changes are usually negligible, Q is positive (adding energy to the fluid in the boiler), thus equation 4.6 simplifies to $\dot{Q} = \dot{m}(e_{h,out} - e_{h,in})$.
- <u>Pumps and compressors</u> are generally used to increase the pressure of the fluid, whilst <u>turbines</u> are used to extract pressure energy. For a Pump/Compressor: $\dot{W}$ is negative, leading to $-\dot{W}$ being positive. Pump power is defined as flow rate × pressure drop. We assume 2 ports, $\dot{Q} = 0$, Potential energy change negligible, Kinetic energy negligible (most of the time), therefore equation 4.6 simplifies to $-\dot{W} = \dot{m}(e_{h,out} - e_{h,in})$. For incompressible fluids density is constant (and not a function of T). Furthermore, when there are relatively small temperature changes in the pump, then $-\dot{W} = \dot{m}/\rho (p_{h,out} - p_{h,in})$.

- Nozzles are devices for increasing the flow velocity at the exit, this might be the end of a hosepipe to a jet engine exhaust. Often the exit pressure and temperature are at atmospheric conditions. Diffusers are the inverse of this, converting kinetic to pressure energy. We assume 2 ports, $\dot{Q} = \dot{W} = 0$, Potential energy negligible. Sometimes you can ignore the inlet KE (as here), then equation 4.6 simplifies to $u_{out}^2/2 = (e_{h,in} - e_{h,out})$. For incompressible fluids density is constant (and not a function of T). Furthermore, when there is a relatively small temperature change, then, a Bernoulli relation is recovered. $\rho u_{out}^2/2 = (p_{in} - p_{out})$.

- Heat exchangers transfer heat from a hot stream to a cold stream, without mixing. Therefore two streams/ four ports, $\dot{W} = \dot{Q} = 0$ (note: no heat transfer across the CV surface), Potential, kinetic energy negligible, then equation 4.6 simplifies to $\dot{m}_1 e_{h,1} + \dot{m}_3 e_{h,3} = \dot{m}_1 e_{h,2} + \dot{m}_3 e_{h,4}$.

- Mixing chambers transfer heat and mass from a hot stream to a cold stream to create a another stream. Here there are three streams, 3 ports, $\dot{W} = \dot{Q} = 0$ (no heat transfer across the control volume surface), Potential, kinetic energy negligible, then equation 4.6 simplifies to $\dot{m}_1 e_{h,1} + \dot{m}_2 e_{h,2} = \dot{m}_3 e_{h,3}$.

	simple form of equation 4.6	assumptions
Burner	$\dot{Q} = \dot{m}(e_{h,out} - e_{h,in})$	$\dot{W} = 0$, 2 ports, $e_g = e_k \approx 0$, $\dot{Q} > 0$
Compressor or Turbine (compressible fluids)	$-\dot{W} = \dot{m}(e_{h,out} - e_{h,in})$	2 ports, $\dot{Q} = 0, e_g = e_k \approx 0$
Pump or Turbine (incompressible fluids)	$-\dot{W} = \frac{\dot{m}}{\rho}(p_{out} - p_{in})$	2 ports, $\dot{Q} = 0, e_g = e_k \approx 0, \Delta T \ll 1$
Nozzle (compressible fluids)	$\frac{u_{out}^2}{2} = (e_{h,in} - e_{h,out})$	2 ports, $\dot{Q} = \dot{W} = 0$, $e_g = e_k \approx 0$
Nozzle (incompressible fluids)	$\frac{\rho u_{out}^2}{2} = (p_{in} - p_{out})$	2 ports, $\dot{Q} = \dot{W} = 0$, $e_g = e_k \approx 0, \Delta T \ll 1$
Heat Exchanger	$\dot{m}_1 e_{h,1} + \dot{m}_3 e_{h,3} =$ $\dot{m}_1 e_{h,2} + \dot{m}_3 e_{h,4}$	4 ports, 2 streams, $\dot{Q} = \dot{W} = 0$, $e_g = e_k \approx 0$
Mixer	$\dot{m}_1 e_{h,1} + \dot{m}_2 e_{h,2} = \dot{m}_3 e_{h,3}$	3 streams, 3 ports, $\dot{Q} = \dot{W} = 0$, $e_g = e_k \approx 0$

Table 4.1: Simplified Forms of the SFEE for Common Engineering Operations

4.8 Conservation of Momentum

As a prelude, remind yourself of scalar and vector quantities (section 2.2.1 on page 9). Then remind yourself of extensive and intensive quantities (section 2.3.3 on page 15), specifically mass (extensive) and its intensive quantity (density – mass per unit volume) are scalars, momentum (extensive) and its intensive quantity (velocity – momentum per unit mass) and force (extensive) and its intensive quantity (acceleration – force per unit mass) are all vectors. So when we write a conservation equation for mass (or density) we write one equation (for instance equation 4.4 on page 63). When we write a conservation equation for momentum (or velocity) we write three equations, conserving momentum in each direction.

We will start by deriving the momentum conservation in one (scalar) direction only. Momentum in the x-direction is $mu_x = M_x$, the momentum vector is $m\vec{u} = \vec{M}$. Rate of change of momentum of a fixed mass is $D/Dt\,(mu_x) = F_x = ma_x$, which shows Newton's second law (equation 2.8 on page 15) is a rate of change of momentum equation. Before we get onto deriving a momentum conservation equation

– remind yourself of the mass conservation equation derivation (section 4.6 on page 63). The key point to note is that the velocity field is transporting mass (density). In what follows we simply apply the same approach, and that the velocity field transports momentum (or in other words, velocity, itself).

First we define specific momentum, $\rho\vec{u}$, momentum per unit volume. By reference to Fig. 4.5, on page 62, the momentum flow in the x-direction through the area element would be $\delta\dot{M}_x = \rho u_x \left(\vec{u}\cdot\vec{n}\right)\delta A$ which has units of $[kg/m^3.m^2/s^2.m^2 = kg.m/s^2 = N]$. Conceptually, the velocity field is carrying a small packet of x-momentum out of the volume. Think again, this is conceptually difficult : the flow does not have to be going in the x-direction, but it carries (convects) some x-momentum. Next, by summing the x-direction momentum flow rate (force) over all the elements, and the net x-direction momentum flow (force) over the entire surface may be defined. $\int \delta\dot{M}_x = \int \rho u_x \left(\vec{u}\cdot\vec{n}\right)\delta A$. This gives us the net x-direction momentum flow (force) through the surface, this must be balanced against one or more x-direction restoring forces, $\sum F_x$ (using Newton's Second Law), otherwise our control volume would move. $\int \rho u_x \left(\vec{u}\cdot\vec{n}\right)\delta A = \sum F_x$. In general, the conservation of momentum is ...

$$\begin{bmatrix} net\ rate\ of\ accumulation \\ of\ momentum\ in\ volume \end{bmatrix} + \begin{bmatrix} net\ rate\ of\ momentum \\ leaving\ the\ volume \end{bmatrix} = \begin{bmatrix} net\ force\ applied \\ to\ surface\ of\ the\ volume \end{bmatrix}$$

If the volume has a volume δV, then the x-momentum in the volume is $\int \rho u_x\, \delta V$, and the rate of change of this is $\partial/\partial t \left(\int \rho u_x\, \delta V\right)$. So the fundamental equation for the conservation of momentum in the x-direction is:

$$\frac{\partial}{\partial t}\left(\int \rho u_x\, \delta V\right) + \int \rho u_x \left(\vec{u}\cdot\vec{n}\right)\delta A = \sum F_x$$

We could write three separate equations, one for each velocity component, or we could write one single vector equation. The general vector form is,

$$\frac{\partial}{\partial t}\left(\int \rho\vec{u}\, \delta V\right) + \int \rho\vec{u} \left(\vec{u}\cdot\vec{n}\right)\delta A = \sum \vec{F} \tag{4.7}$$

For steady flow that has a uniform velocity at each inlet/outlet a simpler version of this equation is $\dot{M}_x = \dot{m}u_x$ and thus for the x-direction $\sum_{out} \dot{M}_x - \sum_{in} \dot{M}_x = \sum F_x$. In addition if there is only two ports and the fluid is incompressible then $\dot{m}(u_{x,out} - u_{x,in}) = \sum F_x$.

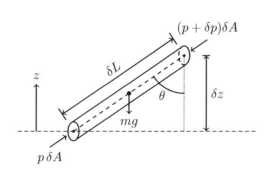

Figure 4.8: A force balance on a stationary fluid element in a gravitational field

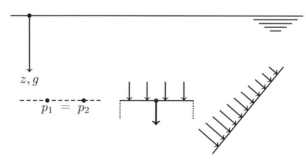

Figure 4.9: Pressure distributions on flat surfaces with normals parallel and at an angle to the gravity direction

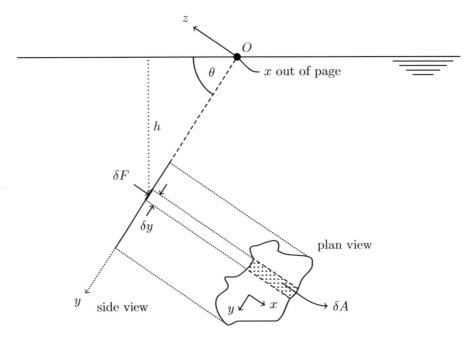

Figure 4.10: *Nomenclature for deriving the force on a flat surface whose normal is not parallel to the gravity vector*

4.9   Force Distributions in Stationary Fluids

The simplest case of a momentum balance is where there is no momentum, and the flow is stationary. Here the forces on the fluid must be in equilibrium, $\sum \vec{F} = 0$ and therefore the pressure forces must balance the weight of the fluid due to gravity.

4.9.1   Hydrostatic Pressure Due to the Weight of the Fluid

Here we relate the pressure change in height to the weight of the liquid in a small cylinder of fluid in order to define the pressure distribution inside fluids. Consider a stationary small cylindrical element of length δL, cross-sectional area δA at an angle θ to the z-direction as shown in Fig. 4.8 where,

- The fluid element is stationary, therefore the forces acting on it must be in equilibrium.
- The weight, defined by the gravity vector acts opposite to the z-direction.
- The pressure force acts in the L-direction, along the fluid element, so we need to resolve the pressure force in the same direction.

The weight acting in the z-direction is $-mg = -\rho V g = -\rho \delta L \delta A g$. In the L-direction the weight is $mg \cos \theta = \rho \delta L \delta A g \cos \theta$. The pressure force acting in the L-direction is $(p + \delta p)\, \delta A - p \delta A = \delta p \delta A$. Equating forces: $-\rho \delta L \delta A g \cos \theta = \delta p \delta A$ and since $\delta z = \delta L \cos \theta$ and in the limit of $\delta z,\ \delta A \to 0$.

$$\frac{dp}{dz} = -\rho g. \tag{4.8}$$

Note the sign convention, here the z and g-directions are opposed, and reflects the pressure reducing with height, for instance in the atmosphere. Regardless of the sign convention a key point is that the pressure is always constant at a constant z-plane if the density is constant as shown in Fig. 4.9. Integration yields, $p = -\int \rho g\, dz$. Finding how p varies with height requires knowing how ρ varies with height. For incompressible fluids $p = -\rho g z + const$, where the constant is defined by the pressure at z=0. Again note the sign convection here follows that of equation 4.8.

4.9.2 Forces on Planar Surfaces Enclosing Fluids

Pressure (more specifically hydrostatic pressure) is force per unit area, $\delta F = p\,\delta A$, in other words a stress. Pressure is a stress magnitude, it is best thought of as acting equally in all directions, and is constant at a given z where the z-axis is parallel to the direction of the gravity. As shown by Fig. 4.9 the normal to a surface defines the direction that the force acts. To work out the pressure force on a surface where the normal to the surface is parallel to the gravity direction is straightforward because the pressure is the same at every point on the surface so $F = \int \delta F = \int p\,\delta A = pA$.

To work out the force on a surface where the surface normal is not parallel to the gravity direction we have to integrate the elemental force distribution along the surface. Figure 4.10 shows a planar stationary surface submerged in a fluid where the surface vertical direction y is at an angle θ to the horizontal fluid surface pressure reference location. The spanwise direction of the surface, x is at constant z, hence pressure and therefore the force on a surface element at a depth h below the fluid surface is $\delta F = p\,\delta A = \rho g h\,\delta A = \rho g y \sin\theta\,\delta A$. The force on the surface is therefore $F = \int \rho g y \sin\theta \delta A = \rho g \sin\theta \int y\,\delta A$. Using the definition of the 1st moment of area (section 2.2.8 on page 13) then $F = \rho g A \overline{h}$ where $\overline{h}$ is the depth of the centre of area below the surface.

To work out where this force can be said to act we have to work out the moment, and we do this using a similar elemental method. As shown in Fig. 4.11 we take moments about point O and the elemental moment in the anti-clockwise direction is defined $\delta M = y\,\delta F = yp\,\delta A = y\,(\rho g h)\,\delta A = \rho g y^2 \sin\theta\,\delta A$. The integral of this must be balanced by the clockwise moment which is defined as the force on the surface multiplied by the point through which this force is said to act, the centre of pressure $F y_p$. Therefore

$$F y_p = \rho g \sin\theta \int y^2\,\delta A \qquad (4.9)$$

thus defining a single location where all the force, distributed over the plate can be said to act.

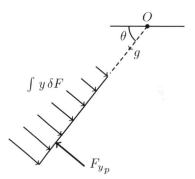

Figure 4.11: *Balancing the element moment on the surface against a single force placed at the centre of pressure y_p*

4.9.3 Forces on Curved Surfaces Enclosing Fluids

Here the typical need is to work out the net force and point of action on a curved surface, for instance the surface A-B in figure 4.12. We can integrate along the gate surface as for planar gates but this requires a coordinate system to be defined for the gate surface. This is certainly possible, but could be very complicated. It is easier to be smarter and sum forces using a free body diagram since this is independent of the shape of the curved surface, as shown by the dashed line insert in Figure 4.12. In effect, we are doing our previous analysis of section 4.9.2 twice, one for each coordinate direction.

Here, we assume a circular gate of unit width as shown in Fig. 4.12. Summing forces in the horizontal x and vertical y directions gives $F_x = F_{AC} = \int p\,dA = \rho g \int y\,dy = \rho g\,[y^2/2] = \rho g (L_2(2L_1 + L_2))/2$. In the vertical direction we have two forces, the weight of the fluid in the body and the force due to the weight of the fluid above it. $F_y = W + F_{CB} = \rho g \pi L_2^2/4 + \rho g L_1 L_2$. The magnitude of the hydrostatic pressure force on the gate is $F = \left(F_x^2 + F_y^2\right)^{1/2}$. We can also define the direction this force acts, $\tan\theta = F_x/F_y$, however we need the centres of pressure in the horizontal and vertical directions to define where this force acts on the gate surface. The centre of pressure for the vertical direction is defined from the total horizontal force and the horizontal moment, much like equation 4.9.

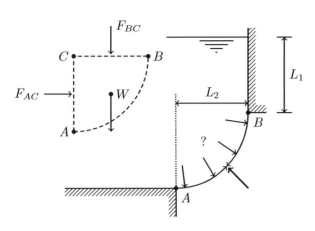

Figure 4.12: *Nomenclature for deriving the force on a curved surface showing control volume used to resolve forces*

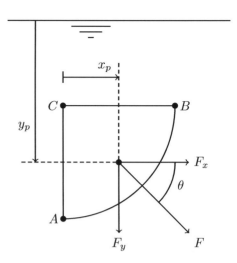

Figure 4.13: *Direction and position of the resultant forces on a curved surface*

$$F_x y_p = \int py \, dA = \rho g \int y^2 \, dy = \rho g \left[\frac{y^3}{3} \right] = \rho g \left[\frac{(L_1 + L_2)^3 - L_1^3}{3} \right].$$

Since $F_x = \rho g \frac{L_2(2L_1+L_2)}{2}$, $y_p = \frac{2}{3} \left[\frac{3L_1^2+3L_2L_1+L_2^2}{(2L_1+L_2)} \right]$. The centre of pressure for the horizontal direction is found by taking moments about C and equating horizontal moments, $F_y x_p = F_{CB} L_2/2 + W \overline{x}$, where $\overline{x}$ is the distance from C to the centre of area of a hemisphere.

For this problem (as noted previously in section 2.2.8 on page 13) the answer is $\overline{x} = 4L_2/3\pi$. Therefore $x_p = 1/F_y(\rho g \pi L_2^3/8 + 4\rho g L_1 L_2^2/3\pi)$. Therefore, as shown by Fig. 4.13, we now have all the information we need. The magnitude and depth at which horizontal force acts F_x, y_p and the magnitude and displacement at which vertical force acts F_y, x_p.

4.9.4 Volumetric (Bouyancy) Forces in Fluids: Archimedes Principle

When rigid bodies are partially or fully submerged in a fluid, two forces are present, the weight of the body, and the upthrust. The latter arises on bodies because of the difference in pressure from top to bottom at a given section. In Fig. 4.14 the pressure distribution around the body is shown. Resolving the vertical components of each of these forces can be found. If we now consider a vertical line (section in 3D, coming out of the page) anywhere on the body from the left hand limit (say) we see two vertical forces are present. The lower one is larger, because the pressure is greater at greater depth. The pressure difference between these two points is the hydrostatic head of fluid between these two points. Integrating the next upward force over surface ADC and subtracting it over the net downward force on surface ABC gives the net upward force, the upthrust. The net upthrust is equal to weight of fluid displaced and is known as Archimedes Principle, $F_{upthrust} = \rho_{fluid} V_{object} g$. Note: although the density of the body itself is immaterial to the amount of upthrust produced, it does of course affect the way in which the body responds to this upthrust.

4.10 Actuator Disk Theory

This is a good example of using simplified forms of the mass (equation 4.4), momentum (equation 4.7) and the energy (Bernoulli, equation 4.3) equations together, and using streamlines and control

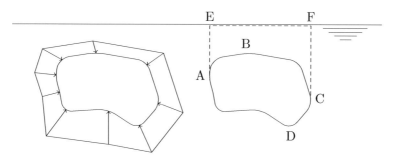

Figure 4.14: *Distribution of force due to hydrostatic pressure on a submerged object*

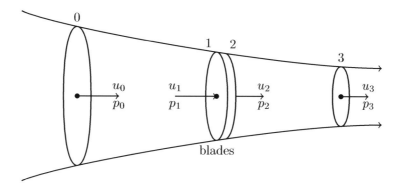

Figure 4.15: *Fluid flowing over a stationary propeller, shown as an actuator disk*

volumes correctly. A propeller, as shown on Fig. 4.16 (on a ship or an aircraft) moves through a stationary fluid, where u is the speed of the object, the "speed of advance" and v is the speed of the oncoming fluid, zero at the moment. The propeller is moving with respect to us, so we change the frame of reference (section 2.3.5.1 on page 17) by moving the air over a stationary propeller. This means, relative to us, the propeller is stationary and sees an advancing air flow of velocity $-v$, speed v. P_{in} is the power input to drive the propeller and transfer occurs by <u>viscous</u> forces acting on the blades of the propeller. Flow through the blades is extremely complicated, and we do not need to know all the complexity from the point of view of working out the power the propeller can impart to/extract from the fluid flow. So we treat it, as shown in Fig. 4.15 as a simple disk shaped <u>control volume</u> into which flow goes in at a certain pressure, and comes out a certain pressure. We assume it is a thin disk of area $A = \pi D^2/4$ and consider a 1-D steady flow of an incompressible fluid.

Considering the control volume (disk), and conserving mass between the fluid at stations 1 and 2, $\dot{m} = \rho A u_1 = \rho A u_2 = const$. Therefore $u_1 = u_2$. Conserving momentum, $(p_1 - p_2)\,A + F_x = \dot{m}\,(u_2 - u_1) = 0$, and note a pressure difference must exist since it generates the thrust (F) for the propeller. We cannot apply the Bernoulli equation here because of the viscous effects on the propeller blade.

We can however apply Bernoulli between stations $0 \to 1$ and $2 \to 3$ because no viscous fforces are present here. Also because stations 0 and 3 are "far" upstream and downstream where we can assume $p_0 = p_3 = p_{atm}$. We can write the Bernoulli equation over stations $0 \to 1$: $\rho u_0^2/2 + p_{atm} = \rho u_1^2/2 + p_1$ and similarly for $2 \to 3$. Since $u_1 = u_2$, these Bernoulli expressions can be added to give a mechanical energy balance from stations $0 \to 3$: $p_2 - p_1 = \rho(u_3^2 - u_0^2)/2$. This gives us an expression for the force, derived from the Bernoulli equations, $F_x = (p_2 - p_1)\,A = \pi D^2/8\rho\,\left(u_3^2 - u_0^2\right)$. Actuator disk theory predicts that $F_x \propto D^2$ and $F_x \propto \left(u_3^2 - u_0^2\right)$.

Now let us examine the velocity difference in more detail. Consider a x-momentum balance over a control volume from $0 \rightarrow 3$, we know no force is applied $0 \rightarrow 1$ and $2 \rightarrow 3$, so the only force present $0 \rightarrow 3$ is the disk. A x-momentum balance over $0 \rightarrow 3$ is: $M_{x,out} - M_{x,in} = F_x$ where $\dot{m} = const$ therefore the thrust generated by the propeller is $F_x = \dot{m}(u_3 - u_0)$. If we substitute $\dot{m} = \pi D^2/4\rho u_1$ and $u_3 = 2u_1 - u_0$ we obtain $F_x = (1/2)\rho\pi D^2 u_1^2 (1 - u_0/u_1)$. Furthermore the power consumed is $F_x u_1$ and the useful power of the propeller is the Thrust × Velocity of advance, $F_x u_0$. This gives the Perfect (lossless) efficiency (Useful/Consumed Power) as $\eta_{perfect} = F_x u_0/F_x u_1 = u_0/u_1$.

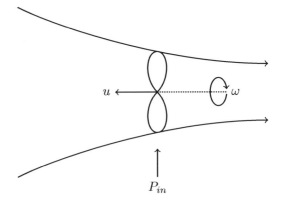

Figure 4.16: *A propellor advancing through still fluid*

Fig. 4.17 shows the theoretical thrust and efficiency characteristics predicted from actuator disk theory. Notice thrust obtained and efficiency are mutually exclusive. The more power you extract via reducing u_0/u_1, the less efficiently you do so. For a given thrust, the bigger you can make your propeller, the more efficient it will be at delivering that thrust. In reality there are losses due to:

- Frictional effects on the propeller blade surface.
- Rotational energy imparted to the fluid when it goes through the propeller.
- Pressure variations on the surfaces of our "black box".

The power losses are (roughly) proportional to the rate of kinetic energy entering the propeller, $P_{LOSS} \propto \dot{m}u_1^2$ or $P_{LOSS} = C\dot{m}u_1^2$, therefore the actual efficiency is

$$\eta_{actual} = \frac{Fu_0}{Fu_1 + P_{LOSS}} = \frac{u_0}{u_1}\left(\frac{1 - u_0/u_1}{1 - u_0/u_1 + C/2}\right).$$

This produces a peak efficiency of 85% for typical aircraft propellers, and is sketched in Fig. 4.17.

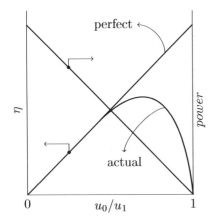

Figure 4.17: *Theoretical Power and Efficiency, and Practical Efficiency of Propellors*

4.11   Jet (Gas Turbine) Engines

Here the aim is to use the Steady Flow Energy Equation (equation 4.5 on page 64) and the Force-Momentum Equation (equation 4.7 on page 67) to describe jet engine operation. An open cycle, control volume description is used, and compared to the Brayton Cycle used for closed cycles (systems) described in section 3.13 on page 54. The basic components of a jet engine are shown on Fig. 4.18.

The inlet diffuser, section $1 \rightarrow 2$ in Fig. 4.18 slows the gas stream down, converting kinetic to pressure energy and then a compressor ($2 \rightarrow 3$) further pressurises the gas. A burner ($3 \rightarrow 4$) adds thermal energy and this is fed into the turbine ($4 \rightarrow 5$) extracts sufficient power to drive the compressor. The nozzle ($5 \rightarrow 6$) converts the remaining pressure energy to kinetic energy. Thrust

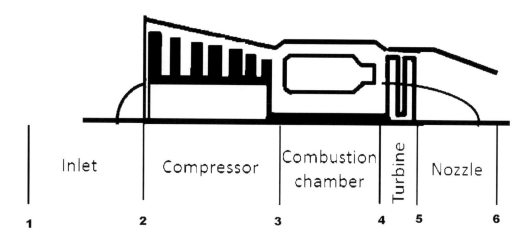

Figure 4.18: *Schematic of Jet Engine Components showing the Diffuser (1-2), Compressor (2-3), Burner (3-4), Turbine (4-5) and Nozzle (5-6)*

(force) developed by the engine is obtained by drawing a control volume around the entire engine and is the difference of momentum in the inlet and the exhaust, $F_x = \dot{m}(u_{out} - u_{in})$. These velocities are relative to the aircraft speed, since in still air u_{in} is the aircraft speed, by changing the frame of reference (section 2.3.5.1 on page 17). The power is the force × distance, i.e. $P = F_x u_{in} = \dot{m}(u_{out} - u_{in})u_{in}$. Finally the efficiency is the useful propulsive power per unit rate of energy input from combustion. $\eta = P/\dot{Q}_{in}$.

4.11.1 Comparing the Jet Engine and the Gas turbine

In the closed Brayton cycle the working fluid goes around the cycle, whilst in the Jet Engine the point is to expel fluid from the exhaust. In the Brayton Cycle the turbine drives the compressor, but the aim of the Brayton Cycle is to maximise turbine output for useful work. In the Jet Engine we want the turbine to be just sufficient to power the compressor since the aim is to preserve as much turbine exit energy for conversion by the nozzle into kinetic energy (thrust). In other words the net work of a jet engine is zero. Finally because some compression occurs in the diffuser section of a jet engine the pressure ratios are generally higher (10-25) than for a Brayton Cycle.

4.11.2 Jet Engine Operational Characteristics

Figure 4.19 shows the $T - s$ diagram for the jet engine open cycle, and points 1 to 6 relate to locations shown on Fig. 4.18. The following process occur:

- Process $1 \rightarrow 2$ Isentropic Compression in a <u>Diffuser</u>.
- Process $2 \rightarrow 3$ Isentropic Compression in a <u>Compressor</u>.
- Process $3 \rightarrow 4$ Constant Pressure heat addition due to combustion in the <u>burner</u>.
- Process $4 \rightarrow 5$ Isentropic Expansion in a <u>Turbine</u>.
- Process $5 \rightarrow 6$ Isentropic Expansion in a <u>Nozzle</u>.
- Process $6 \rightarrow 1$ (Theoretical) – this is the energy lost due to low temperature gases being ingested by the engine and high temperature gases being expelled.

We now instead of drawing a control volume around the entire engine, draw control volumes around the individual components of the engine listed above. We then apply the SFEE in simplified form, from table 4.1 on page 66. In a diffuser the area increases, and the fluid slows down, and vice versa for the nozzle. Following section 4.7.3 on page 65 the diffuser energy balance is $0 = e_{h,out} - e_{h,in} + u_{in}^2/2$.

In the compressor/turbine work is added/extracted to raise/lower the pressure energy. Following section 4.7.3 the compressor energy balance is $-\dot{W} = \dot{m}\left(e_{h,out} - e_{h,in}\right)$. In the combustion section heat energy is added, this is $\dot{Q} = \dot{m}\left(e_{h,out} - e_{h,in}\right)$. In terms of efficiency $P = Fu_{in} = \dot{m}\left(u_{out} - u_{in}\right)u_{in}$ and typical efficiencies are $\eta = P/\dot{Q}_{in} \sim 22\%$. The two primary losses are the excess kinetic energy of the exhaust gas, relative to the ground, $\sim33\%$ and the excess enthalpy, the temperature rise of the exhaust gas over ambient air, $\sim45\%$. Or put another way for every 100kg of fuel burnt only 22kg is used to get you from A to B.

4.12   Key Points

We now have a basic understanding of how to conserve mass, momentum and energy over a fixed volume.

- Pathlines and streaklines are flow traces over time, whereas streamlines are a representation of the flow at all locations at a single point in time.
- The Euler Equation defines the force balance along a streamline and the Bernoulli equation a mechanical energy balance.
- The Steady Flow Energy Equation conserves total energy in a fixed volume
- The Steady Mechanical Energy Equation is a mechanical energy balance in a fixed volume, treating the thermal terms as a pressure drop.
- The momentum change in a certain direction must be balanced by a force in the same direction.
- A force is said to act at the centre of pressure of a surface in a hydrostatic pressure field.
- Forces on curved surfaces are obtained using a free body diagram.
- The upthrust is the weight of fluid displaced.

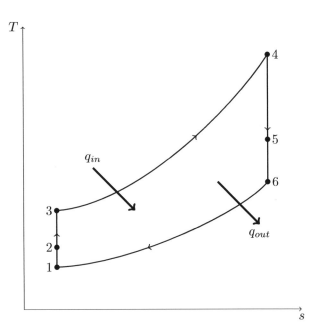

Figure 4.19: *T-s diagram of the Jet Engine Processes*

5 | Introduction to Viscous Flow

We now consider processes where *both* viscous (diffusive) and inertial (convective) forces are important and here we restrict ourselves to incompressible fluids and build on the more simple introduction to simultaneous convection and diffusion of heat in section 2.6 on page 23. Viscous forces are usually important near walls and control the stability of boundary layers and transition to turbulence, so the chapter is really about flow near walls. These issues are of tremendous importance in terms of understanding aircraft flight and drag reduction in ships, pipes, blood vessels for instance. The key initial knowledge required for this chapter is the diffusion of momentum and the Newtonian Stress-Strain relationship for a fluid, defined by equation 2.13 and discussed in section 2.6.3 on page 24. This entire chapter is about the impact of this little equation.

5.1 The Nature of Turbulence

Turbulence is partially a viscous process and is the norm rather than exception in the environment, in science and in engineering. The dynamics of clouds are a classic example, the smoke rising from a cigarette another. An important property made use of by engineers is to increase the rate of mixing, we all know that milk 'diffuses' more quickly in a cup of tea if we stir it. A disadvantage is that energy is lost due to the turbulent motion, and for instance it takes a bigger pump to transport a turbulent fluid than a non-turbulent (laminar) fluid. Turbulence has some very specific characteristics that differentiate it from fluid flows that are simply unsteady or periodic.

- <u>Large Reynolds Numbers</u> : Turbulence is only possible when the inertial forces exceed the viscous forces of the flow, and this is defined by a *large* Reynolds Number. This is case specific, for instance pipe flows are normally turbulent when the Re~2300.
- <u>Dispersive and Three Dimensional</u> : Turbulence always spreads information in all directions, it is dispersive, like the milk being spread throughout the volume of the tea cup.
- <u>A Range of Eddy Sizes</u> : Usually turbulence is generated by the flow passing over an object and that object creating a turbulent wake, a good example is a the flow visualised by a flag attached to a cylindrical flag pole. As the fluid moves over the flag pole large eddies, of the scale of the flag pole diameter are shed. We see them as the flapping of the flag. These eddies entrain more fluid, twist, break up and over time the energy of the turbulence (the kinetic energy of the fluctuations) transfers from the large eddies to progressively smaller eddies. A key problem for engineers is that the range of eddy scales (from large to small) increases very quickly with Reynolds Number. Clouds have enormous Reynolds Numbers, and the largest eddies might by $\sim 1km$ in size the smallest might be $\sim 1mm$, making weather prediction even more difficult.
- <u>Dissipation</u> : Continuing our flow over the flag pole illustration, we can imagine the 'turbulence energy' is created at the large eddy scales (of the scale of the flag pole) and is transferred to smaller and smaller eddies. Eventually the viscous forces become so large that the viscous force and the flow velocity at these small scales converts the kinetic energy into heat. This is an irreversible process which creates entropy and is an important cause for energy losses due to turbulence in process equipment. Following on from this, because turbulence is dissipative, kinetic energy must be continually supplied to maintain it.

A poem by LF Richardson summarises the last two points thus :

Big whorls have little whorls
That feed on their velocity,
And little whorls have lesser whorls
And so on to viscosity.

5.2   Steady Viscous Diffusion of Momentum: Couette Flow

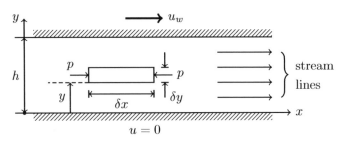

Figure 5.1: *A control volume in a Couette Flow*

Figure 5.1 shows two very large plates, normal to the y-axis, with nothing changing in the x and z-directions. The top plate is moving with a velocity u_w and the bottom plate is stationary. There is no fluid acceleration in the x-direction, the flow is fully developed, and therefore, no pressure changes in this direction. There is no fluid motion in the y-direction and no pressure changes in the y-direction since we neglect hydrostatic force here. The only force the fluid feels is due to molecular diffusion of x-momentum transferred from the moving plate into the fluid *in the y direction*. We take an elemental volume, δx by δy by unit width (into the page) and we assume we know what the stress is per unit width at y: $\tau|_y \delta x = \mu du/dy|_y\, \delta x$. We use a Taylor series to define the stress at $y + \delta y$:

$$\tau|_{y+\delta y}\, \delta x = \left\{ \tau|_y + \delta y \frac{d}{dy}\,\tau|_y \right\} \delta x = \left\{ \mu \frac{du}{dy}\bigg|_y + \delta y \frac{d}{dy} \frac{du}{dy}\bigg|_y \right\} \delta x$$

and since there are no other forces on the control volume shown in Fig. 5.1 then $\tau|_y = \tau|_{y+\delta y}$ or $d^2u/dy^2 = 0$ since the control volume is stationary and the net force on it must be zero. If we integrate this twice we get $u(y) = Ay + B$ and then apply boundary conditions ($u = 0$ at $y = 0$ gives $B = 0$ etc.) to give $u(y) = u_w y/h$. Since the stress is defined by equation 2.13 on page 25 then $\tau = \mu u_w/h$. There is a linear variation in the velocity due to constant shear stress because of constant viscosity. This shows momentum in the x-direction diffusing in the y-direction from the moving wall to the stationary wall.

5.3   Rate of Viscous Diffusion of Momentum: Stoke's First Problem

Couette Flow (section 5.2) demonstrates the steady state diffusion of momentum, whilst Stoke's First Problem demonstrates the rate at which momentum diffuses into the fluid. It examines the rate at which momentum diffuses into the fluid from a wall that is initially stationary and then impulsively started to a velocity U. As a first approximation it can be thought of as the near wall start up condition, for instance for the moving wall of Couette Flow, above. The partial differential equation solved is,

$$\frac{\partial u}{\partial t} = \nu \frac{\partial^2 u}{\partial y^2} \tag{5.1}$$

where for $t > 0$, $u(y = 0, t) = U$, in other words the velocity of the wall at $y = 0$ goes from $u = 0$ to $u = U$ at $t = 0$. From $t > 0$ momentum starts to diffuse into the fluid from the wall, as shown by Fig. 5.2, which shows the velocity profile in the fluid at several times after $t = 0$. The solution to equation 5.1 is $u(y,t)/U = 1 - erf\left(\frac{y}{2\sqrt{\nu t}}\right)$ where $erf(x)$ is the error function.

The point is the form of the non-dimensional term inside the error function. This relates the change in the viscosity to the rate at which the velocity at a certain y and t reaches a certain value. In other words how quickly momentum diffuses into the liquid. We see that $t \propto \nu$, double the viscosity and you half the time at which a certain amount of momentum reaches a certain height above the wall in the fluid. Clearly the viscosity controls the rate at which momentum diffuses into the fluid in the the same way thermal controls the rate at which heat diffuses through a material at a given temperature gradient.

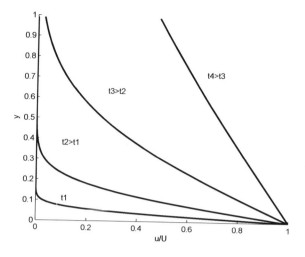

Figure 5.2: *Evolution of the velocity profile above an impulsively started wall*

5.4 Viscous Diffusion of Momentum: Fully Developed Pressure Driven (Poiseulle) Pipe Flow

In a pressure driven flow of a viscous fluid the pressure drop, $dp/dx = (p_1 - p_2)/(x_2 - x_1) = -\Delta p/L$. This must be $-ve$ as p must fall with x to be balanced by the shear stresses at the wall τ_w. This is shown graphically in Fig. 5.3, for steady fully developed flow in the x-direction (therefore nothing depends on x, including dp/dx). The key point is the force due to the pressure gradient is balanced by the shear stress at the wall. The average (bulk mean) velocity is $\bar{u}$, the volume flow rate, $\dot{V} = \bar{u}A = \bar{u}\pi R^2$ and the power required to move the liquid is $\dot{V}\Delta p$. If we define a force balance on the control volume shown in Fig. 5.3 of length L then $(p + \Delta p)\pi R^2 - p\pi R^2 - \tau_w 2\pi R L = 0$ or $\Delta p/L = 2\tau_w/R$ or

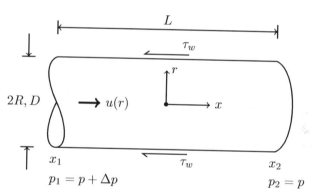

Figure 5.3: *A control volume defined by the the fluid inside a a section of pipe*

$$\tau_w = -\frac{R}{2}\frac{dp}{dx} \qquad (5.2)$$

This is a fundamental result because it does not depend on the state of the flow, providing it is fully developed. It relates the wall shear stress to the pressure drop. We now perform the same analysis, but now we take a shell of fluid at $R > r > 0$ and small length δx, as shown in Fig. 5.4. This time we prescribe the pressure change in the x-direction in terms of a Taylor Series (note that $dp/dx < 0$), therefore

$$\left(p - \left(p + \frac{dp}{dx}\delta x\right)\right)\pi r^2 - \tau 2\pi r\,\delta x = 0.$$

This time we assume that the flow is laminar, hence assume equation 2.13 on page 25 holds, y is the distance from the wall, here $y = R - r$ and $dy/dr = -1$, therefore $\tau = \mu du/dy = \mu(du/dr)(dr/dy) =$

$-\mu du/dr$. Then substitute shear stress definition (equation 2.13) into the force balance to give $du/dr = (r/2\mu)dp/dx$. This is integrated and a wall boundary condition (section 2.6.4.3 on page 26) is applied to obtain the velocity profile,

$$u(r) = -\frac{R^2}{4\mu}\frac{dp}{dx}\left(1-\left(\frac{r}{R}\right)^2\right) \tag{5.3}$$

and the shear stress distribution.

$$\tau(r) = -\mu\frac{du}{dr} = -\frac{1}{2}\frac{dp}{dx}r \tag{5.4}$$

It can be seen that setting $r = R$ in eqn. 5.4 recovers eqn. 5.2, however it is important to remember that eqn. 5.2 is a general equation for any flow, at the wall whereas eqn. 5.4 is only valid for laminar flow, but for any r.

Equation 5.3 shows that the velocity profile in a laminar fully developed round pipe is parabolic whilst equation 5.4 shows the shear stress distribution is linear. Both of these are shown in Fig. 5.5. The stress distribution linearly increases from zero on the centreline and the wall shear expression (equation 5.2) is recovered at the wall. A more common form of the velocity profile is $u/u_{\max} = \left(1-(r/R)^2\right)$. The volume flow rate can be found the integral of the elemental flow rate, $\delta\dot{V} = u2\pi r\,\delta r$,

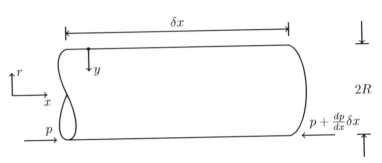

Figure 5.4: *Nomenclature for a control volume element in the pipe*

$$\dot{V} = 2\pi\int_0^R ur\,dr = -\frac{\pi R^4}{8\mu}\frac{dp}{dx}.$$

Since the bulk mean flow $\bar{u} = \dot{V}/A$ then $\bar{u} = u_{\max}/2$. We can also related the bulk mean flow directly to the pressure drop,

$$\bar{u} = -\frac{R^2}{8\mu}\frac{dp}{dx} \tag{5.5}$$

5.4.1 Coefficient of Friction and the Moody Diagram

From an engineers point of view, for a given pipe diameter, flow velocity, fluid characteristics, we need to know the pressure drop per unit length of pipe, in order to size a pump. We can pull all the previous information together to develop a dimensionless "friction coefficient". Using dimensional analysis (section 2.4) we might expect that $\tau_w = f(\rho, \mu, \bar{u}, D)$, and since we know the wall shear defines the pressure drop per unit length for any flow (section 5.4), knowing this is sufficient to size our pump.

This leads to $\tau_w/(1/2\rho\bar{u}^2) = f(\rho\bar{u}d/\mu) = f(Re)$, where $d = 2R$, the pipe diameter. Note this is a functional relationship and the LHS has already introduced in eqn. 5.8 on page 81. It is the C_f term and it is known as the "friction factor". Other names include the "Fanning Friction Factor", and the "Darcy Friction Factor", the latter equal to $C_f/4$ and is used by Chemical Engineers. Clearly dimensional analysis tells us that C_f depends on the pipe Reynolds Number is some way. It is also completely general in the sense

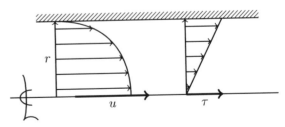

Figure 5.5: *Laminar Velocity and Shear Stress Profiles in Fully Developed Laminar Pipe Flow*

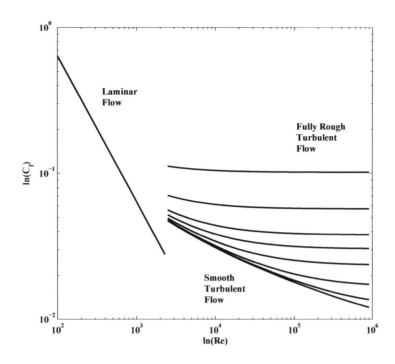

Figure 5.6: *The Moody Diagram - Pipe Friction Coefficient versus Reynolds Number*

that no physics have been assumed, therefore it applies
to laminar and to turbulent flow. For a laminar flow assumption and using the general definition for
wall shear, τ_w in eqn. 5.2, we know the velocity gradient at the wall from the laminar flow mean
velocity (equation 5.3) we can write a specific relation for the laminar flow friction coefficient.

$$C_f = \frac{\tau_w}{1/2\rho\overline{u}^2} = \frac{16}{\mathrm{Re}}. \tag{5.6}$$

The laminar flow friction factor does not depend on the roughness, as shown in Fig. 5.6, which is
known as the Moody Diagram. On the vertical axis is C_f. Also note the laminar friction factor fails at
the critical Reynolds number for pipe flow, $\sim$2100. Beyond the critical Re the pipe flow is turbulent
and the wall shear/pressure drop has three relationships (1) where the roughness is so small that that
pressure drop only depends on Re (2) where both the roughness and Re influence C_f and finally (3)
where C_f is independent of Re and depends only on the roughness. Note also that although the
laminar C_f is greater than the turbulent C_f because the shear stress at the wall (from equation 5.6)
is $\tau_w = (1/2)C_f\rho\overline{u}^2$, in other words the shear stress rises very steeply with the bulk mean flow, much
faster than the drop shown on the Moody diagram for increasing Re.

5.5   The Steady Mechanical Energy Equation (SMEE)

A conceptual simplification to the SFEE (equation 4.6 on page 64) would be to only consider mechanical energies, and this is what the steady mechanical energy equation (SMEE) represents. This,
conceptually is what the Bernoulli equation does too, however because it is based on a streamline care
must be applied when applying it. The SMEE, because it is based on the same control volume basis
as the SFEE has no such restrictions and as will be shown can simplify to the Bernoulli equation. A
typical application for the SMEE is for estimating pressure drops in pipe networks, therefore the first
thing we do is simplify the steady flow energy equation (equation 4.5) to only have two ports (inlet
and outlet) and that the only important velocity component, u, flows parallel to the pipe axis, such

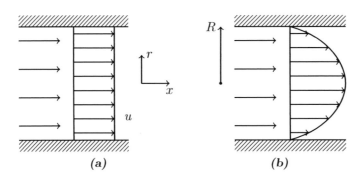

Figure 5.7: *Uniform and Non-Uniform pipe velocity profiles*

that,

$$\dot{Q} - \dot{W} = \int_{A,out} \left(C_v T + u^2/2 + p/\rho + gz\right) \rho u \, \delta A - \int_{A,in} \left(C_v T + u^2/2 + p/\rho + gz\right) \rho u \, \delta A.$$

We know that $\dot{m} = \int_A \rho u \, \delta A$, and that $\dot{m}_{out} = \dot{m}_{in}$ so most of these terms simplify:

$$\dot{Q} - \dot{W} = \dot{m} \left(C_v T + p/\rho + gz\right)_{out} + 1/2 \int_{A,out} u^2 \rho u \, \delta A$$

$$- \dot{m} \left(C_v T + p/\rho + gz\right)_{in} - 1/2 \int_{A,in} u^2 \rho u \, \delta A. \tag{5.7}$$

We need to analyse the kinetic energy term a little more to simplify that. If the velocity along the point was uniform (ie u did not depend on the pipe radial direction) then the KE would simply be $\dot{m} u^2/2$. This is only true if the fluid is inviscid, which is never true in practice, and viscous forces are ALWAYS important near a wall. This is the main reason you CANNOT use the Bernoulli Equation to work out the pressure drop in a pipe (or anything else with a wall). Fig. 5.7a shows a uniform (inviscid fluid) velocity profile and Fig. 5.7b shows a real (viscous fluid) velocity profile, for the same bulk mass flow $\dot{m} = \int \rho u \, \delta A = \rho \bar{u} A$. Although the mass flow is the same in Fig. 5.7a and Fig. 5.7b the kinetic energy flow, $\int u^2/2 \rho u \, \delta A$ is different , KE(a) $= \dot{m} u^2/2$ while KE(b) $= 1/2 \int_0^R u^3 \, dA$. A KE correction factor is defined such that $\alpha = KE(b)/KE(a)$, so that $KE(b) = \alpha KE(a) = \alpha \dot{m} \bar{u}^2/2$. Note the use of the bulk mean velocity (volume flow rate/area). The KE correction factor has the following values, $\alpha = 1$ (uniform flow), $\alpha = 2$ (laminar flow), $2 > \alpha > 1$ (turbulent flow, typically 1.1 for fully turbulent flow). It is important to note it must be applied to each flow boundary individually, because the pipe inlet might be laminar and the pipe exit might be turbulent. Equation 5.7 is now simplified to

$$\frac{\dot{Q}}{\dot{m}} - \frac{\dot{W}}{\dot{m}} = \left(C_v T + p/\rho + gz + 1/2\alpha \vec{u}^2\right)_{out} - \left(C_v T + p/\rho + gz + 1/2\alpha \vec{u}^2\right)_{in}.$$

It should be noted that other than steady flow and 2 ports, no other assumptions have been made. The steady mechanical energy equation is designed to be an isothermal viscous flow equation, so we re-arrange the equation to group the thermal components.

$$-\frac{\dot{W}}{\dot{m}} = \left(p/\rho + gz + 1/2\alpha \vec{u}^2\right)_{out} - \left(p/\rho + gz + 1/2\alpha \vec{u}^2\right)_{in} + \left[C_v \left(T_{out} - T_{in}\right) - \frac{\dot{Q}}{\dot{m}}\right].$$

The thermal terms in the square brackets on the RHS represent a rate of mechanical energy loss. They are an irreversible temperature rise due to viscous dissipation, and the transfer of that thermal energy through the pipe wall, $\dot{Q}$. If the pipe is lagged $\dot{Q} = 0$ obviously. This energy loss is real and significant and must be considered in design calculations for viscous flows near walls and through engineering plant. The thermal terms as presented here are not practically useful, empirical models are used instead, expressing the loss as a pressure drop. To give some idea of the magnitude, we can relate the temperature rise to a pressure drop using the SFEE again, e.g. $C_v T_1 + p_1/\rho = C_v T_2 + p_2/\rho$ or $\Delta p = \rho C_v \Delta T$. For a 1K temperature for water say ($\rho \sim 10^3 kg m^{-3}, C \sim 10^3 J kg^{-1} K^{-1}$), $\Delta p \sim 10^6 N m^{-2}$, or 10 bar.

5.5.1 Models for Viscous Losses in the Steady Mechanical Energy Equation (SMEE)
The viscous dissipation term is usually expressed in terms of a kinetic energy change and two types of energy losses (i.e. causing a change in pipe pressure) are typically defined, wall friction and fittings, the first and second terms of the right hand side of the equation below. The losses ALL arise due to viscous dissipation of kinetic energy.

$$[C_v (T_{out} - T_{in})] \Rightarrow C_f \frac{L}{d} \left(1/2\overline{u}^2\right) + \sum_n k_n \left(1/2\overline{u}^2\right).$$

Wall friction requires a coefficient, C_f, for a pipe of length L and diameter d. For laminar flow this may be derived analytically and this is done in section 5.4.1 on page 78. For turbulent flow an empirical chart known as a Moody diagram must be used, as is discussed more fully in section 5.4.1 on page 78. Losses due to fittings, for instance due to bends, T-junctions, expansions and contractions are defined by setting the k_n factors empirically, and a good source is Perry's Chemical Engineers Handbook. This gives the final form of the steady mechanical energy equation.

$$-\frac{\dot{W}}{\dot{m}} = \left(p/\rho + gz + 1/2\alpha\vec{u}^2\right)_{out} - \left(p/\rho + gz + 1/2\alpha\vec{u}^2\right)_{in} + C_f \frac{L}{d} \left(1/2\overline{u}^2\right) + \sum_n k_n \left(1/2\overline{u}^2\right). \quad (5.8)$$

Often when using this equation to size a pump for instance one must solve the problem iteratively, guessing and correcting the velocity value and Reynolds Number and hence C_f.

5.5.2 Beware the "Extended Bernoulli Equation"
Note you will find in several textbooks equation 5.8 is often referred to as the "Extended Bernoulli Equation". It is NOT and cannot be derived from the Bernoulli Equation. Starting from the steady mechanical energy equation (equation 5.7), if we assume: (a) No work transfers (b) No viscous losses (c) uniform flow we can then simplify to the Bernoulli Equation (equation 4.3). Even though it is mathematically identical, it STILL isn't exactly the same, because one (equation 5.7) was derived for a finite volume of fluid while the other (equation 4.3) is derived for a point fluid element moving along a streamline. You can go from equation 5.7 to equation 4.3 but you CANNOT start from equation 4.3 and derive equation 5.7 because the basic assumption of the Bernoulli equation is that the fluid is inviscid and the loss terms of the extended Bernoulli equation *rely* on viscosity.

5.6 Structure of a Fully Developed Turbulent Boundary Layer

In section 5.1 on page 75 a brief explanation of turbulent flow was given, and the key characteristics were that a range of different sized eddies interact with each other and large eddies tend to twist and break up into smaller ones. There is also a limit in terms of how small turbulent eddies can be before viscosity destroys them and the kinetic energy of the turbulent motion is dissipated as heat.

When a wall is present in a turbulent flow, the structure of the turbulence is influenced by the presence of the wall. The distance from the wall is determined in terms of non-dimensional viscous wall units, using a variable y^+. When $y^+ \sim 1$, one viscous wall unit, viscous effects are of the same order as inertial effects. Another way of thinking about y^+ is a wall Reynolds number, with the distance away from the wall the length scale. Because of this, next to the wall there is a very thin layer where the viscous forces are too large and turbulent eddies, of a scale similar to the width of this "inner" layer

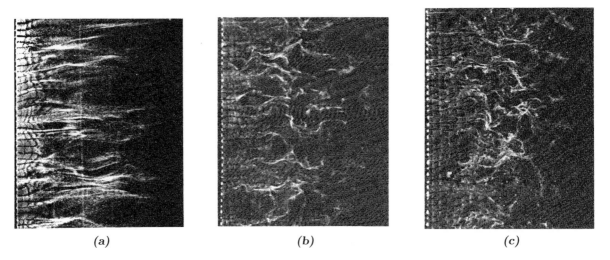

Figure 5.8: *The structure of eddies above a flat plate at, from L to R, $y^+ = 2.7, 38, 100$*

cannot exist. So, right next to the wall the flow is, on average, laminar. A little further away from the wall, in a region called the "buffer layer", where $y^+ \sim 10$ turbulent eddies are present, which act to diffuse momentum around in a similar way that viscous momentum transfer does. Viscous momentum forces are still important here. Finally in an "outer" region, where $y^+ > 50$ the flow is dominated by the turbulent motion and viscous forces have little contribution to transporting momentum.

The shape of the turbulent eddies is also a function of the wall distance. As shown by Fig. 5.8, where $y^+ \sim 3$, the turbulence is very "streaky" and runs along the wall surface, further out, say $y^+ \sim 40$, these streaks become convoluted and by the time we get into the fully turbulent region, say $y^+ \sim 100$, then the turbulent eddies have lost all memory of their origins near the wall. Each of these eddies interact with each other and the wall in a different way and produce an individual contribution to the wall shear. Clearly this makes a simple analytical approach to estimating the pressure drop in turbulent pipe flow like we did for laminar flow near impossible.

Instead we resort to experiments, plot the data in terms of the correct dimensionless groups and curve fit the results. There are two reasons why this is important (1) most industrial pipe flows are turbulent and (2) the pressure drops for turbulent flows are larger than for laminar flow. This is not surprising, all that turbulent energy must come from somewhere, and it is all dissipated as heat eventually. To see how important this issue is we compare the laminar and turbulent pressure drop per unit length expressions. For laminar flows we can use equations 5.2, 5.5 and 5.6 to define the pressure drop per unit length.

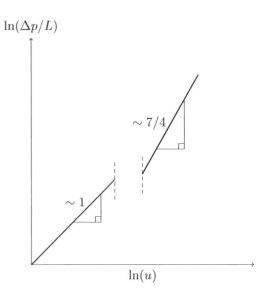

Figure 5.9: *Pressure Drop per Unit Length due to Laminar and Turbulent flow*

$$\frac{\Delta p}{L} = \frac{2}{R} 1/2 \rho \bar{u}^2 \frac{16}{\text{Re}} = \frac{8\mu}{R^2} \bar{u} = g\left(\bar{u}^1\right) \qquad (5.9)$$

For turbulent pipe flow Prandtl proposed $C_f = 0.079\,\text{Re}^{-1/4}$ which gives.

$$\frac{\Delta p}{L} = \frac{0.079}{R} \rho \bar{u}^2 Re^{-1/4} = g\left(\bar{u}^{7/4}\right) \qquad (5.10)$$

Taking logs of equations 5.9 and 5.10, as shown on Fig. 5.9, we can see the pressure drop increases at a much faster rate in turbulent flow than it does in laminar flow. This is bad news most engineering pipe flows, which are turbulent and hence need bigger pumps.

5.6.1 Effect of Roughness on pressure drop in turbulent boundary layers

The final piece of the puzzle in terms of the influence turbulent boundary layers have on pressure drop in a pipe is the influence of length scale of the roughness, κ as shown on the Moody diagram. This may be understood by defining the roughness in terms of the viscous wall distance unit y^+ mentioned in section 5.6. We can define a Reynolds number based on distance from the wall, $\text{Re}_y = yu/\nu$ (which is really the same thing as y^+) and we would expect viscous effects to be important when $\text{Re}_y \sim 1$, and so $y = \nu/u$. So when $\kappa \ll y$ then we can consider the wall "smooth". All walls have roughness, it just depends how big the roughness is compared to the thickness of the laminar inner layer. In the other limit, when the $\kappa \gg y$ then the roughness disrupts the normal streaky structures on the wall surface and creates eddies by vortex shedding. Because this is nothing to with viscosity, therefore there is no Reynolds number dependence for the pressure drop for very rough walls. Also for larger u, y reduces and smaller roughness has the same effect. These characteristics are shown on the Moody diagram, Fig. 5.6 on page 79.

5.7 Developing Boundary Layers

In section 5.4 we discussed laminar boundary layers and in section 5.6 the characteristics of turbulent turbulent boundary layers. In both of these cases we made the assumption that the flow was _fully developed_, that the state of the boundary layer did _not_ depend on the position along the pipe. Obviously, when the flow enters a pipe, it takes some time, or distance along the pipe for the boundary layer to develop fully. In this section we analyse _developing_ boundary layers. First, internal boundary layers are briefly introduced then, in more detail, external boundary layers are discussed.

5.7.1 Internal Developing Boundary Layers

At the entrance to a pipe as shown in Fig. 5.10, a boundary layer will initiate and develop, its height at a given point is the boundary layer thickness and usually given the symbol δ. The region of fluid unaffected by the boundary layer in the developing region is known as the potential core. Eventually the boundary layers from the walls will meet and then the flow will be fully developed. Note that the wall shear stress (equation 5.4 for $r = R$ and laminar flow) is initially high near the entrance due to the thinness of the boundary layer (ie the velocity changes by a large amount over a small distance), and reduces as it thickens. At fully developed conditions (by definition) the wall shear is not a function of position along the pipe.

5.7.2 External Boundary Layers on Flat Surfaces

Here we consider "external" boundary layers, the evolving flow near a solid surface, a classic example is the flow over an aircraft wing. The key difference between internal and external boundary layers is that external ones evolve along the surface they are caused by, they are never "fully developed". Outside the boundary layer the fluid is undisturbed and has the velocity of the free stream. It can often be treated as inviscid, thus the viscous shear force, and hence drag to shear stresses occurs solely in the boundary layer. At the wall a no-slip boundary condition applies – the relative velocity of the fluid with respect to the wall is zero. The edge of the boundary layer is usually taken to be 99% of the free stream (relative) velocity, other possible boundary conditions are listed in section 2.6.4 on page 25. External boundary layers may be laminar or turbulent and may transition from the former to the

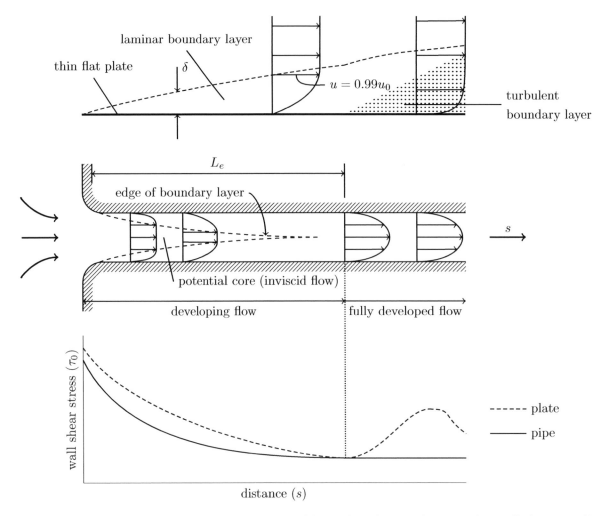

Figure 5.10: *Developing external and internal boundary layers showing the wall shear profile*

latter. Figure 5.10 shows the key features of an external boundary layer, here a thin flat plate placed edge on in a uniform flow. A laminar boundary layer of thickness δ initiates at the leading edge and the shear stress as the wall is initially very high because the velocity changes from the free stream to zero in a very short distance. As the boundary layer thickens the wall shear stress reduces. At some point the boundary layer may "trip" and convert to a turbulent boundary layer. At this point the boundary layer "starts again" and because it is thin the turbulent shear stress is large. Thereafter the turbulent boundary layer develops and the wall shear stress again reduces.

5.7.3   Properties of Laminar versus Turbulent Boundary Layers

Laminar and turbulent boundary layers have different properties that are exploited in different applications. A good example is an aircraft wing. As may have been surmised from section 5.6, laminar boundary layers have lower wall shear stresses than turbulent boundary layers, because du/dy is larger. This is known as friction drag. Because turbulent boundary layers have more energy in them, they are less likely to separate, which can happen when the surface turns away from the flow, as detailed in section 5.7.5 in full. The properties of the boundary layer can be chosen to suit the application. For instance aerofoils need to ensure the flow on the upper surface does not separate during take off and landing, even when the angle of attack is quite steep, so a turbulent boundary layer at this location is preferred. Once airborne at cruise however the primary requirement is to save fuel and reduce drag, so a laminar boundary layer is preferred. For good control, both laminar and turbulent boundary layers are present, and the the transition point is controlled by altering the geometry of the wing.

5.7.4 Displacement and Momentum Thicknesses

Because boundary layers are difficult to deal with, and as engineers we are usually more interested in the effect of the boundary layer on mass or momentum changes to the flow, equivalent displacement (mass) and momentum thicknesses of boundary layers are used. Fig. 5.11a shows a typical boundary layer velocity profile near a wall and we could if we want to work out the mass flow in the shaded area. In Fig. 5.11b we have moved the wall into the fluid by a distance δ^*, which corre-

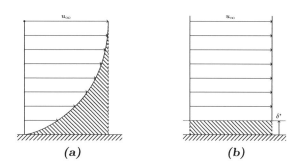

Figure 5.11: *Displacement Thickness of a Boundary Layer*

sponds to the mass flow shaded in Fig. 5.11b. Equating mass defects (unit width): $\rho \int_0^\infty (u_\infty - u)\, dy = \rho u_\infty \delta^*$ or $\delta^* = \int_0^\infty (1 - u/u_\infty)\, dy$. We can also equate the momentum defect, which gives a different thickness θ. $\rho \int_0^\infty u_\infty (u_\infty - u)\, dy = \rho u_\infty^2 \theta$ or $\theta = \int_0^\infty u/u_\infty (1 - u/u_\infty)\, dy$. Their ratio is known as the boundary layer "shape factor", $H = \delta^*/\theta$ and $H = 2.59$ for a laminar boundary layer and 1.3-1.6 for a turbulent boundary layer. This is useful because for instance the drag force on a flat plate can be defined directly in terms of the free stream velocity and the momentum thickness, $D = \rho u_\infty^2 \theta$.

5.7.5 External Boundary Layers on Curved Surfaces

We have until now only examined the wall shear stress distributions (and hence the <u>friction</u> drag) for flat surfaces, Couette flow (section 5.2 on page 76), Laminar (section 5.4) and turbulent (section 5.6) pipe flows and developing boundary layers on flat plates (section 5.7.2). Flat plates have zero pressure gradients along them, there is no separation, no pressure (or form) drag. However we know from our work on dimensional analysis that the drag relationship is complex, even for simple shapes like cylinders, and we know this relationship is Reynolds number dependent. First we define form drag, then examine how adverse pressure gradients lead to separation.

5.7.5.1 Pressure (form) drag

When a surface element of an object is part of the boundary layer, if that surface element is not parallel to the flow, a normal force will exist in addition to the familiar shear force (friction drag) discussed in section 2.4.4. Figure 5.12 shows an elemental force balance on a surface element inclined at an angle θ to the free stream flow. The elemental drag force is $dF = (-p\cos\theta + \tau\sin\theta)\, dA$. The first part is known as the pressure or form drag, the second the familiar friction drag.

5.7.5.2 Adverse Pressure Gradients and Separation

We take a streamlined body as shown in Fig. 5.13 and trace a streamline going around that body. If we applied the Bernoulli equation to that streamline then we would see the pressure decrease as the fluid is accelerated around the nose section of the aerofoil and $dp/ds < 0$. Beyond the point of maximum thickness of the body the pressure increases as the fluid and $dp/ds > 0$. In the latter case the pressure gradient is considered adverse, and promotes separation. Boundary layer separation occurs when the local velocity gradient at the wall tends to zero. From equation 5.4, so does the shear stress. In terms of the total friction drag change on the surface, the change may not be dramatic, however the wake induced by this boundary layer separation may be catastrophic to both the drag and the lift performance of the object. Turbulent boundary layers tend to be better at delaying the separation point, and so although they produce high friction drag, for certain objects the wake is smaller and there is less *form* (or pressure) drag and hence less drag overall. This is why golf balls have dimples.

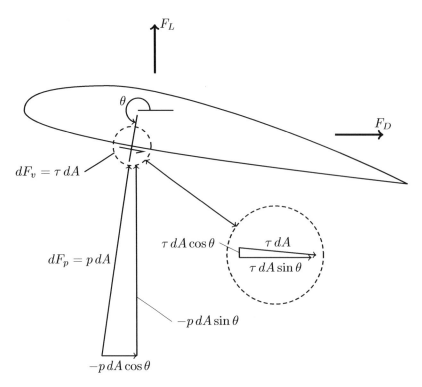

Figure 5.12: *A section of a Wing Surface Showing both Pressure (form) and Viscous (friction) Drag.*

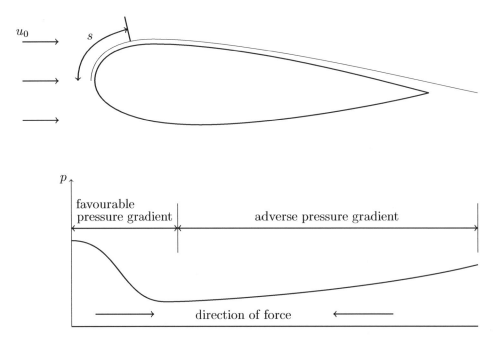

Figure 5.13: *Pressure Change along a streamline near the top surface of a streamlined object*

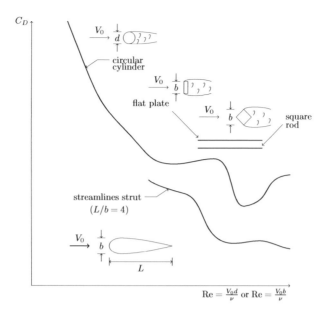

Figure 5.14: *Drag Co-efficient for different shapes as a function of Re*

Figure 5.14 shows the drag coefficient, $C_D = D/(A\left(1/2\rho u^2\right))$ for a variety of shapes as a function of the object Reynolds number. For rounded objects like the cylinder the linear section is purely friction drag, and there is no separation. As Re increases a wake forms and form drag component rises and finally the boundary layer goes turbulent on Re $\sim 10^5$ and the drag suddenly drops as the wake contracts as the separation point moves further round the downstream surface of the cylinder. Bluff bodies are defined as having a sharp edge which causes separation of the flow independent of the Reynolds Number, hence the independence of drag with Reynolds number for these bodies. Streamlined objects by design have low drag coefficients, and generally smoother drag curves with the separation point gradually moving along the surface as the Reynolds Number changes.

5.8   Key Points

We now have a basic understanding of how momentum diffuses in viscous flow, and how boundary layers can form, transition and separate. We have covered,

- Viscosity controls the rate at which momentum in the x-direction diffuses into the fluid in the y-direction
- Couette Flow is a classic momentum diffusion problem and Poisuelle (pipe) flow is the viscous wall shear balanced by the pressure drop along the pipe.
- Pressure Drop in a pipe is defined by the Moody Diagram which shows the effect of Reynolds Number and Wall Roughness.
- The probability of flow separation rises in external boundary layers when the pressure gradient goes to zero or is positive.
- Turbulent boundary layers produce more friction drag but are less prone to separation, a key producer of form (pressure) drag.

6 | Vector and Cartesian Tensor Notation

In section 2.2.1 on page 9 scalars and vectors were introduced, and one vector operator, the dot product. This minimal approach was taken to ensure we could get the conceptual message across, without getting bogged down in overly detailed, but correct, definitions. For instance equation 2.13 on page 25 defines the shear stress, as a scalar, in terms of a velocity gradient. In reality (ie in 3D) this is only partially correct. The aim of this chapter is to introduce common mathematical notation systems used in continuum mechanics simply but formally. We need to do this to derive the full 3D conservation equations, of which the conservation of momentum (equation 4.7 on page 67) and the steady flow energy equation (equation 4.5 on page 64) are simplifications of. Vector and Cartesian tensor notation is the language of mathematical science and engineering and once you grasp it, you will see the systematic way in which stress, strain, force, pressure and motion inter-relate. Once you get to this level of awareness, continuum mechanics becomes very easy, since you can see the pattern in which physical information relates to itself. In this chapter we are going to:

1. To introduce the concept of scalar, vector and tensor quantities, from a physical point of view.
2. To introduce vector and Cartesian tensor "notation", a way to represent these physical quantities in a clear logical and compact way that aids their analysis and manipulation.
3. To introduce vector and Cartesian tensor "operators", ways to manipulate and transform physical quantities, defined in vector/Cartesian tensor form, into other physical quantities.
4. To permit the learning of continuum mechanics to focus on the physical basis of the equations, rather than their method of notation.

6.1 Organisation of this Chapter

This is an introduction to the subject of vector and Cartesian tensor notation and operations on these vector and Cartesian tensor 'objects'. The subject has a mathematical basis, so definitions are usually very precise, and sometimes quite involved, and it can be difficult to initially find the important bits of information. The aim here is to focus on the main message, and try to not get lost in the mathematical detail. On the other hand, ultimately it only makes sense with the detail, so here the body of the text will contain the main message, and the detail will be relegated to footnotes. It is recommended the first couple of times you go through this, you simply ignore this detail. However, once you have read through a few times and you see enough of the big picture, go back through the footnotes and other supplementary material to *understand* how this mathematical notation system all hangs together.

6.2 Basic Terminology

In this section we set the mathematical scene we are going to explore, so to do that we need to define some terms and introduce some concepts, which may or may not be familiar. Here we are going to introduce two different ways to represent scalar, vector and tensor quantities, they are known as 'vector' notation and 'Cartesian tensor' notation. Their properties are summarised in table 6.1.

Vector notation has its roots in geometrical calculation, and can be easily understood by relating vector operations to trigonometric methods. Tensor notation is defined through the transformation of information from one coordinate system to another. Because of this, and the requirement that a

Property	Vector Notation	Cartesian Tensor Notation
Primary use	Define variables, operations and equations compactly and independent of the coordinate system	Define variables, operations and equations compactly in the Cartesian coordinate system
Advantages	What is written is completely general and universal to any coordinate system	All variables and operations are defined directly within the nomenclature
Disadvantages	To undertake operations, you have to define the coordinate system and memorise or look up the operations. Some operations are ambiguous. Nomenclature systems vary	You are restricted to operations in the Cartesian coordinate system.

Table 6.1: *Vector and Tensor Notation Comparison*

physical law must be independent of the coordinate system used in describing them mathematically, tensor notation is the correct way to represent and manipulate equations.

Unfortunately, people and textbooks use these systems interchangeably across science and engineering, so to understand the information in these texts, you need to do likewise. My recommendation is to use Cartesian tensor notation if you can, because the notation fully defines the operations at a component level and properly defines equations expressing physical laws. This becomes self-evident as you read through this document, providing you know a few basic rules. Right now, you only need to know one.

6.2.1 Rules of Cartesian Tensor Notation (rule 1 only)

The "rank" of the 'tensor' is defined by the number of 'free' indices. We need to define the words in quotes, and the use of them is shown throughout the document. Right now, please just accept that:

1 A 'tensor' is a type of mathematical object to group information which has the same mathematical and in many cases the same physical characteristics.
2 The 'rank' (of the tensor) defines the number of dimensions the object has.
3 The rank is defined by the number of 'free' indices is the number of non-repeated subscripts present on the tensor.

For instance, a tensor with two free subscripts is a 2-rank tensor, an example is the stress tensor, σ_{ij}. Another would be Z_{ijkk}. Physically, an item of information in a 2-rank tensor is defined by two directions. With this in mind, let us proceed.

6.3   Variables, Points, Fields, and Gradients of Variables

A variable is some property you are interested in, like temperature, pressure or force. Points and fields (of information, of a physical property, like temperature) define the spatial extent of the information. A point variable (temperature say) defines the value of that variable at that location in space[1]. A field

---

[1]More specifically, defines the variable at a given location in a given coordinate system.

variable defines the values of the variable in a number of dimensions along a line (1D), on a surface (2D), or in a volume (3D). Conceptually, think of a field variable as an array of point variables.

The gradient of a variable is the rate of change of that variable with respect to a coordinate direction. For instance if $y = x^2$ then the gradient of y with respect to the x coordinate direction is differential of y with respect to x is $2x$. Just as a variable can be a point or a field, so too can a gradient. To calculate a gradient however, you need information from more than one *point*. In other words you need a *field* of information to get a gradient *point* or a *field*.

The physical reason why gradient fields are so important in continuum mechanics that they are spread information through diffusion processes, as introduced in section 2.6.2 on page 23. The conduction of heat through a solid, the diffusion of a chemical species through a still fluid, the diffusion of momentum in a viscous moving fluid are all defined by the relevant gradient field. For instance the bigger the gradient field of temperature the bigger the flux[2] of heat in the direction of that gradient[3]. Vector and tensor notation allow a compact notation for this flux.

Example 1

For some variable f a one dimensional field is represented, in a Cartesian coordinate system, by $f(x)$, two dimensional by $f(x, y)$ and three dimensional by $f(x, y, z)$[4]. An example of a 3D scalar field would be:

$$T(x, y, z) = x^3 y - z^2 \qquad (6.1)$$

Example 2

Let us say we have a one dimensional steel bar (in the x-direction) of length L, heated at one end (at $x = 0$) cooled at the other ($x = L$), and perfectly insulated along its length. At some *point* the temperature is T, and the temperature *field* in the entire bar is $T(x)$. Let us say that the temperature field is defined by the equation

$$T(x) = T_{hot} - \frac{(T_{hot} - T_{cold})\, x}{L}$$

To calculate the temperature *gradient* field, in the x-direction, we differentiate the temperature field with respect to x, i.e.

$$\frac{d}{dx}\{T(x)\} = -\frac{(T_{hot} - T_{cold})}{L}$$

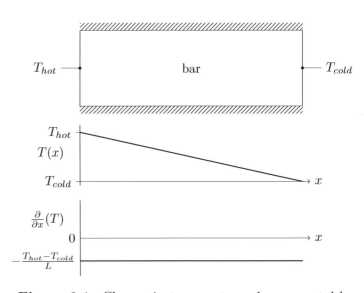

Figure 6.1: *Change in temperature along a metal bar*

Visually, you can see:

---

[2] A flux is a flow per unit area. For instance the heat flux is the heat flow (J/s) per unit area, and has units of $J/m^2 s$. Flux is a vector quantity, containing both direction and magnitude information

[3] More generally, almost all continuum mechanics systems have "diffusion" present, and the strength of the diffusion is defined by two things, the magnitude of the gradient, and a diffusion coefficient. For thermal energy the coefficient is the thermal conductivity, for chemical species it is a diffusion coefficient, for momentum it is the viscosity.

[4] Herein lies the main use for vector and tensor notation: to represent and manipulate 3D fields of information in a compact and systematic way. Note very often the (x) or (x, y) notation for field quantities is dropped.

- The temperature reduces linearly from $x = 0$ to $x = L$.
- Because of this the gradient of the temperature field is constant.
- Since the gradient represents the heat flux[5] (heat flow per unit area) and the area of the bar is constant, then the heat flow along the bar is also constant. Since the bar is perfectly insulated, providing the thermal conductivity is constant, this has to be correct.

6.4 Scalars, Vectors and Tensors[6]

Continuum[7] mechanics share many physical quantities, such are pressure, temperature, speed, velocity, force, acceleration, strain, and stress. Many of these physical quantities have similar *mathematical* characteristics too and are defined in terms of the *type* of information they contain. Here we define three *types* of information, *scalars*, *vectors* and *tensors* and they differ in terms of the *dimensionality* (or rank) of the information they contain.

6.4.1 Scalars (0-rank tensor)
- Mathematically, a scalar is something that only has one element of information.
- Conceptually a scalar is a zero dimensional information "container".
- Physically (i.e. in engineering) this item of information is an amount or a size or more precisely a *magnitude*. For example the scalar pressure at a point, P, is the magnitude of the scalar pressure field at that point.

Examples include

- Energy (i.e. thermal and kinetic energy).
- Pressure, Temperature
- Speed (not velocity).
- Length or distance (not position or displacement).
- Most physical properties (viscosity, thermal conductivity, specific heat etc.).
- Magnitudes of vectors are scalars (see below).

In both vector and Cartesian tensor notation, scalars are represented as you would normally represent a point or a field variable, for instance T, $T(x)$, $T(x, y, z)$ etc.

6.4.2 Vectors (1-rank tensor)
- Mathematically, a vector is something that has three elements of information and is represented by a 3 element row (1×3) or column array (3×1) matrix.
- Conceptually a vector is a one dimensional information "container".
- Physically (ie in engineering) these three elements always relate to coordinate directions (for instance (x, y, z) in a 3D Cartesian coordinate system, and represent quantities that have direction information. For instance the vector position defines the x, y and z-displacements of a point from the co-ordinate system origin.

---

[5]Assuming the thermal conductivity is also constant.

[6]Strictly, everything is a tensor: a "scalar" is a zero rank tensor, a "vector" is a first rank tensor, and a "tensor" is a second rank tensor. The term "rank" implies the dimensionality of the information in the tensor being defined.

[7]Fluid mechanics and solid mechanics are both forms of continuum mechanics. Solids require the understanding of force, pressure, stress and strain, and relating these physical concepts to material properties in order to understand the strength of materials and structures. Fluid mechanics requires this understanding, and also the concept of convection, in order to understand how fluids move and the types of forces required to create fluid motion, and conversely the forces exerted on structures by fluid motion, for instance wind loading on buildings.

Examples include

- Position
- Velocity (not speed).
- Acceleration, Force, Momentum.
- Gradients of scalar quantities (the gradient of the [scalar] temperature field is a vector).
- Heat flux
- Vorticity[8] (a rotation vector)

The way a vector is written down depends on the type of notation used. There are two forms in common use and both have their advantages and disadvantages (see below). In your studies you will be expected to understand information described with either method, so here they are presented side by side.

Property	Vector Notation[9]	Cartesian Tensor Notation
Vector (velocity)	$\vec{u}$	u_i, u_j

Table 6.2: *Vector and Tensor Notation for a Rank 1 Tensor*

In Cartesian Tensor notation the i and j subscripts can take values 1, 2, 3, corresponding to the x, y and z-coordinate directions[10]. A velocity vector can be represented 3D Cartesian coordinate system as one rank (dimension) of information: in row form, $u_i = \begin{pmatrix} u_x & u_y & u_z \end{pmatrix} = \begin{pmatrix} u_1 & u_2 & u_3 \end{pmatrix}$ in column form[11].

$$u_j = \begin{pmatrix} u_x \\ u_y \\ u_z \end{pmatrix} = \begin{pmatrix} u_1 \\ u_2 \\ u_3 \end{pmatrix}$$

You do not need to know why the row and column forms exist right now, just please accept they do exist, and that the index can flip from i to j, and it doesn't change what the vector (velocity say) actually is. Note both the i and the j index both still represent the x-y-z coordinates mapped to values 1-2-3 (axes x, y, z). If you have a 3-D scalar field, then you can find the gradient of that field, in the x-direction say, at every point in that field. You can then repeat the gradient calculation in the y and z-directions. Therefore you can generate a *vector* gradient field from a scalar field. An important additional property a vector contains is the vector magnitude, which is a scalar quantity. The magnitude is represented as follows:

Property	Vector Notation[12]	Cartesian Tensor Notation
Vector Magnitude (speed)	$|\vec{u}|$	$|u_i|$

Table 6.3: *Vector and Tensor Notation for tensor magnitude*

---

[8]Vorticity is a "pseudovector" but not a Cartesian tensor. This is explained more fully in Appendix A and B of Stephen Pope "Turbulent Flows".

[9]Vector notation in some texts is represented as bold, for example A is a scalar and **B** is a vector. Here we use an arrow overhead, $\vec{B}$.

[10]Because the x, y and z-coordinate directions relate to the 1, 2 and 3 coordinate directions, we usually write x as x_1, y as x_2, z as x_3

[11]There is also a k form. normal to the plane of the page.

[12]Vector notation in many texts is represented as bold, for example A is a scalar and **B** is a vector. Here we use a single underscore, ie B to ensure the vector definition is clear, regardless of the how this document is viewed.

Therefore vectors contain both direction and magnitude information. From a purely geometrical perspective, a vector can be thought of as an arrow, with the vector magnitude representing the length of the arrow. A vector *field* can be thought of as an array of arrows.

Example 3

The *velocity* vector $u_i = (3 \quad 4 \quad 0)$ at position $x_i = (1 \quad 1 \quad 0)$ defines a direction and magnitude of the flow at the given (vector) position. The magnitude of the (velocity) vector (the speed) can be obtained by trigonometry (a 3 4 5 triangle) and $|U_i| = \left(U_1^2 + U_2^2 + U_3^2\right)^{1/2} = 5$. Furthermore if you divide each vector element, by the (scalar) vector magnitude you define a unit vector. This is a vector which has unit magnitude (length) but the same direction information as the original. If you use vectors to define direction only, it is common to use these unit vectors[13].

Property[14]	Vector Notation	Cartesian Tensor Notation
Unit Vector	$\hat{\vec{u}}$	$\hat{u}_i$

Table 6.4: Vector and Tensor Notation for Unit Tensors

Example 4

$$\hat{u}_i = \frac{u_i}{|u_i|} = \begin{pmatrix} \frac{3}{5} & \frac{4}{5} & 0 \end{pmatrix}$$

and by definition $|\hat{u}_i| = 1$.

Example 5

Let us imagine we have a steel cube of side L, aligned with the Cartesian coordinate system axis, with one corner of the cube placed at the origin of the coordinate system. It is heated uniformly throughout the volume, and faces normal to y and z axes are perfectly insulated, and the faces normal to the x-axis are maintained at a temperature T_C. The 3D temperature distribution is $T(x, y, z) = T_C + C\left(x^2 - Lx\right)$. Note that the temperature field is a function of x only. In other words "the temperature only changes with respect to x". So if we differentiate with respect to x, then y and then z we obtain three equations. $dT/dx = dT/dx_1 = 2Cx - L$, $dT/dy = dT/dx_2 = 0$ and $dT/dz = dT/dx_3 = 0$. Like example 2, the gradient field tells us the direction and the magnitude of the heat from the volume (where heat is being generated) to the walls. We can see that:

- There is no heat flow in the 2 and 3 directions (because walls normal to these surfaces are insulated, so no heat flows in this direction.
- There is heat flow in the 1 direction, symmetrical about $x = L/2$. Heat flow is zero at $x = L/2$ and increases linearly away from this plane (because more heat has to carried through the solid as you get nearer the wall).

The three one dimensional gradients are scalar quantities, and together they are the three components of the *vector* gradient field of the scalar temperature. The nomenclature for the vector gradient field (of a scalar) is as follows:

Property	Vector Notation	Cartesian Tensor Notation

---

[13]This is especially true to define the (vector) direction of each coordinate system axis, for instance the x-axis of an x-y-z coordinate system would be *(1,0,0)*.

[14]This nomenclature is not universal

Gradient of a scalar field	$\vec{\nabla}T$		$\partial T/\partial x_i$

Table 6.5: *Vector and Tensor Notation for the gradient operator*

In Cartesian tensor notation the gradient field notation maps directly to the components of the vector like any other vector. In this example,

$$\frac{\partial T}{\partial x_i} = \begin{pmatrix} \frac{\partial T}{\partial x_1} & \frac{\partial T}{\partial x_2} & \frac{\partial T}{\partial x_3} \end{pmatrix} = \begin{pmatrix} 2Cx - L & 0 & 0 \end{pmatrix}$$

Finally, before we look at tensors, the general case is the gradient field of a n-rank variable has rank $n+1$. So in this example, a temperature field is a scalar, a 0-rank tensor and the gradient field of a scalar is a vector, a 1-rank tensor. This is so because for each temperature point (ie at a specific x, y, z location) we can work out a gradient in the x, y and z-directions. These three gradient components, grouped together, define the *vector* temperature gradient field.

6.4.3   Tensors (2-rank tensor)

2-rank tensors (or just 'tensors') can be thought of several ways, contain a lot of information and can get quite complicated. Therefore, at this stage, we are just trying to define (a) what a 2-rank tensor is and (b) how a 2-rank tensor fits into the 0-1-2 rank tensor system of notation.

- Mathematically tensors contain 9 elements of information, arranged in a 3×3 matrix.
- Conceptually a tensor is a two dimensional information "container".
- Physically tensors represent information that needs to be defined by two coordinate directions.

Again the way a tensor is written down depends on the type of notation used. First we try to illustrate

Property	Vector Notation	Cartesian Tensor Notation
Tensor (stress)	$\vec{\vec{\sigma}}$	σ_{ij}

Table 6.6: *Vector and Tensor Notation for rank 2 Tensors*

what a tensor is and here we limit the discussion to a one physical example, the stress tensor. It is defined in terms of the 9 components, arranged in a 3×3 matrix.

$$\sigma_{ij} = \begin{pmatrix} \sigma_{xx} & \sigma_{xy} & \sigma_{xz} \\ \sigma_{yx} & \sigma_{yy} & \sigma_{yz} \\ \sigma_{zx} & \sigma_{zy} & \sigma_{xz} \end{pmatrix} = \begin{pmatrix} \sigma_{11} & \sigma_{12} & \sigma_{13} \\ \sigma_{21} & \sigma_{22} & \sigma_{23} \\ \sigma_{31} & \sigma_{32} & \sigma_{33} \end{pmatrix}$$

The tensor has 2 indices (or ranks or dimensions) i and j, and define the location of the tensor element in the matrix. i defines the horizontal increment and j the vertical increment, in exactly the same way as row and column *vectors* are defined using the same notation[15]. Both i and j can take value of 1, 2 or 3, which again correspond to the coordinate directions x, y and z.

The stress tensor is used to represent the stress exerted on a surface. The first index defines the dimension (direction) is the direction the stress is acting. The second index defines the surface normal.

---

[15]More generally, the first index defines the horizontal direction and the second index defines the vertical increment. Like vectors, the indices do not have to be i and j, and they do not necessarily be in order, in other words ij and both ji are valid. The distinction, however, can be important.

For instance σ_{12} defines the stress acting in the 1 (or x) direction due to a physical process acting on a surface with a normal in the 2 (y) direction.

When the stress and the surface normal direction are the same, the stress tensor component is said to be a *normal component*, and physically this relates to a squeezing or stretching processes.

When the stress and the surface normal direction are different, the stress tensor component is said to be *shearing component*, and physically this relates to deformation and rotational processes.

Normal stress components are defined in the diagonal of the 3×3 matrix, one for each coordinate direction. Shear stress components are defined in the off-diagonal locations in the 3×3 matrix. Now we try to illustrate how 2-rank tensors are part of the pattern going from scalars (0-rank tensors), vectors (1-rank tensors) to tensors (2-rank tensors). We have already shown how a gradient field of a scalar field (a 0-rank tensor) is a vector field (a 1-rank tensor). Now let us look at the gradient field of a vector field (1-rank tensor), which we show is a tensor field (2-rank tensor).

Example 6

In example 5 we looked at how a gradient vector field can be generated from a scalar field. Now we do the same for a gradient field of a vector field. The gradient field of a (1-rank) vector field is a (2-rank) tensor field. Imagine our cube in example 5 is now a box, mostly filled with fluid, and it is shaken for a while. At the instant the shaking is stopped the velocity field, a *vector field*, will be non uniform. Because the velocity field is non-uniform, velocity *gradients* will exist, and this example will show how all this velocity gradient information can be "packaged" into a 2-rank tensor. Let us represent our velocity field in Cartesian tensor form, $u_i\,(x, y, z)$. This means at every (x,y,z) location in the box, we have three bits of information, the velocity components in the x, y and z-directions, u_x, u_y, u_z, or, in Cartesian tensor notation, u_1, u_2, u_3. Remember, each velocity component *field* is a scalar *field*. To make things similar to example 5, let us take the u_1 velocity *field*, e.g. $u_1\,(x, y, z)$, and remember, this is a scalar field. Now, read through example 5 again, and satisfy yourself that $u_1\,(x, y, z)$ and $T\,(x, y, z)$ are from a mathematical point of view, exactly the same thing. So, we can write the gradient field of the *scalar* u_1 velocity component field as,

$$\frac{\partial u_1}{\partial x_j} = \begin{pmatrix} \frac{\partial u_1}{\partial x_1} & \frac{\partial u_1}{\partial x_2} & \frac{\partial u_1}{\partial x_3} \end{pmatrix}.$$

Here we are using a j index, which can take values of 1, 2, 3 to represent the gradient of the u_1 velocity in the x, y and z-direction. So for this velocity component, there exists the gradient of this velocity component in the three coordinate directions. Now, we did not have to use u_1, we could have used the other two velocity components, so we can easily write for the other two *scalar* velocity components.

$$\frac{\partial u_2}{\partial x_j} = \begin{pmatrix} \frac{\partial u_2}{\partial x_1} & \frac{\partial u_2}{\partial x_2} & \frac{\partial u_2}{\partial x_3} \end{pmatrix} \quad \text{and} \quad \frac{\partial u_3}{\partial x_j} = \begin{pmatrix} \frac{\partial u_3}{\partial x_1} & \frac{\partial u_3}{\partial x_2} & \frac{\partial u_3}{\partial x_3} \end{pmatrix}.$$

Now, we have three velocity components, with three gradient components each, which makes 9 components. Each of these 9 terms has two bits of direction information, for instance, $\partial u_3 / \partial x_2$ defines the gradient of the $u_3 (u_z)$ velocity component in the $2(y)$ direction. Tensors (i.e. 2-rank tensors) can represent this type of information, so we can write the velocity gradient *tensor*,

$$\frac{\partial u_i}{\partial x_j} = \begin{pmatrix} \frac{\partial u}{\partial x} & \frac{\partial u}{\partial y} & \frac{\partial u}{\partial z} \\ \frac{\partial v}{\partial x} & \frac{\partial v}{\partial y} & \frac{\partial v}{\partial z} \\ \frac{\partial w}{\partial x} & \frac{\partial w}{\partial y} & \frac{\partial w}{\partial z} \end{pmatrix}$$

where $u_i = (u_x \quad u_y \quad u_z) = (u_1 \quad u_2 \quad u_3) = (u \quad v \quad w)$ which can be written as

$$\frac{\partial u_i}{\partial x_j} = \begin{pmatrix} \frac{\partial u_x}{\partial x_x} & \frac{\partial u_x}{\partial x_y} & \frac{\partial u_x}{\partial x_z} \\ \frac{\partial u_y}{\partial x_x} & \frac{\partial u_y}{\partial x_y} & \frac{\partial u_y}{\partial x_z} \\ \frac{\partial u_z}{\partial x_x} & \frac{\partial u_z}{\partial x_y} & \frac{\partial u_z}{\partial x_z} \end{pmatrix} = \begin{pmatrix} \frac{\partial u_1}{\partial x_1} & \frac{\partial u_1}{\partial x_2} & \frac{\partial u_1}{\partial x_3} \\ \frac{\partial u_2}{\partial x_1} & \frac{\partial u_2}{\partial x_2} & \frac{\partial u_2}{\partial x_3} \\ \frac{\partial u_3}{\partial x_1} & \frac{\partial u_3}{\partial x_2} & \frac{\partial u_3}{\partial x_3} \end{pmatrix}$$

6.4.4   Special Tensors

There are two special tensors that we need to define. Here we just state what they are, in Popes Appendices, you can see why they exist and how they are used.

6.4.4.1   Kroneker delta tensor

This is a 2-rank tensor, where the normal components of that tensor are unity, and the off-diagonal elements are zero. It has a special symbol δ_{ij}, and may be stated,

$$\delta_{ij} = \begin{pmatrix} 1 & 0 & 0 \\ 0 & 1 & 0 \\ 0 & 0 & 1 \end{pmatrix}$$

6.4.4.2   Alternating unit tensor[16]

This is a 3-rank tensor, and has 27 elements, arranged in a $3 \times 3 \times 3$ matrix, and has the symbol ϵ_{ijk}. These 27 elements contain the values +1,0,-1, depending on the following rules

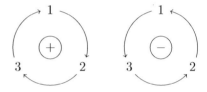

Figure 6.2: *Sign convention for the alternating unit tensor*

1 If the i, j and k indices are different, and increasing, cyclically from left to right then the element value is +1. For example $\epsilon_{123} = \epsilon_{231} = \epsilon_{312} = 1$.

2 If the i, j and k indices are different, and decreasing, cyclically from left to right then the element value is -1. For example $\epsilon_{321} = \epsilon_{213} = \epsilon_{132} = -1$.

3 If any two indices are the same the value is 0. For example $\epsilon_{331} = \epsilon_{223} = \epsilon_{112} = 0$.

An easy way to remember this is as shown in Figure 6.2 with the 1, 2, 3 numbers arranged clockwise around a circle. Clockwise sequences are positive, anti-clockwise sequences are negative.

6.5   Rules of Cartesian Tensor Notation

The key to understanding what type of Cartesian tensor is present and what is occurring in a tensor operation is understanding what is meant when the Cartesian tensor subscripts change. You do this by following these simple rules, the first of which you have already been given.

1 The "rank" of the "tensor" is defined by the number of "free" indices.
2 No index may be repeated more than once.
3 If you have a repeated index, then you must sum the components along that rank.
4 In a tensor equation, all terms in that equation must have the same number of free indices, and the free indices used must all be the same.

---

[16]This is another "pseudotensor", see Pope, Appendix A,B for details.

Example 7

A, B_{ii}, B_{iijj} are all 0-rank tensors (scalars) because they all have 0 free (non-repeated) indices. They have one component of information.

A_i, B_j, B_{iik} are all 1-rank tensors (vectors) because they all have 1 free index. They each have three elements of information, for instance the velocity vector.

A_{ij}, B_{jk}, B_{iijk} are all 2-rank tensors (tensors) because they all have 2 free indices. They all have 9 elements of information.

A_{iii}, B_{jjjk} are all invalid tensors, because an index is repeated more than twice.

$A_i = B_i$, $A - B = 0$, $A_{ij} = B_{ijkk} + C_{ij}$ are all valid tensor equations, because the free indices are consistent in each term in of the equation.

$A_i = B_j$, $A_i - B = 0$, $A_{ik} = B_{ijkk} + C_{ijk}$ are all invalid tensor equations.

A repeated index in a tensor implies a tensor operation, called a "tensor contraction". First let us introduce some[17] tensor operations, with some physical examples in each case. Theses operations apply to both point tensors and also field tensors. Later, we will discuss some operations that only apply to fields (we have already met one, the gradient operator).

6.6 Cartesian Tensor Operations[18]

Now we know what different types of tensors are, we can now start to learn how they can be used, and again, for continuum mechanics, here I show the physical significance of performing these *operations*. Here we define operations which you already know about for ordinary scalar variables, like addition, subtraction and multiplication and to this list we add some new operations specific to tensors of rank higher than 0. In this, and all subsequent sections of this document, the operations at a component level, will be defined using Cartesian tensor notation, because there is an explicit correspondence between the nomenclature of the operation, and what you have to actually do at the component level. Also, if you already know about matrix operations, then Cartesian tensor operations are very very similar. Vector operation nomenclature will be presented alongside the tensor operation in table form as in the section above, so if you see these operations, in vector form, in text books, then you will understand what is being said.

6.6.1 Addition and Subtraction of Tensors

You can add and subtract a tensor of the same rank[19] in just the same way as you add and subtract a scalar, i.e. $4 - 3 = 1$. When you do an addition/subtraction operation on a tensor, you add/subtract the same components from each tensor to make a new tensor of the *same* rank. This operation is valid for tensors of all ranks.

Property	Vector Notation	Cartesian Tensor Notation
Addition/Subtraction (here showing a 2-rank tensor addition operation)	$\vec{C} = \vec{A} + \vec{B}$	$C_{ij} = A_{ij} + B_{ij}$

Table 6.7: *Addition/Subtraction of Tensors*

Example 8

---

[17] Not all are listed.

[18] 'operation' here means 'do something'. For instance adding two tensors together is an operation on these two tensors.

[19] More specifically a tensor with the same free indices.

Two vectors are defined $A_i = (1 \quad 2 \quad 3)$ and $B_i = (3 \quad 2 \quad 1)$, the addition of these two vectors is written as $C_i = A_i + B_i$. At the component level then $C_i = (C_1 \quad C_2 \quad C_3) = (A_1 + B_1 \quad A_2 + B_2 \quad A_3 + B_3) = (4 \quad 4 \quad 4)$.

6.6.2   Scalar Multiplication and Scalar Division of a Tensor

Here we multiply a tensor of any rank by a scalar. This operation does not change the rank of the tensor, but does change the magnitude of each component in the tensor. In the table below, C is a scalar and A and B are tensors. The scalar multiplication operation requires each element of the

Property	Vector Notation	Cartesian Tensor Notation
Scalar Multiplication (of a 2-rank tensor)	$\vec{B} = C\vec{A}$	$B_{ij} = C A_{ij}$

Table 6.8: Scalar Multiplication of a Tensor

tensor to be multiplied by the scalar constant. Note you cannot divide by a tensor of order 1 or more, so therefore, strictly, vector force is *defined* by $F_i = pA_i$.

Example 9

A shear stress tensor is $\sigma_{ij} = \begin{pmatrix} \sigma_{11} & \sigma_{12} & \sigma_{13} \\ \sigma_{21} & \sigma_{22} & \sigma_{23} \\ \sigma_{31} & \sigma_{32} & \sigma_{33} \end{pmatrix} = \begin{pmatrix} 1 & 2 & 3 \\ 4 & 5 & 6 \\ 7 & 8 & 9 \end{pmatrix}$, so $4\sigma_{ij} = \begin{pmatrix} 4 & 8 & 12 \\ 16 & 20 & 24 \\ 28 & 32 & 36 \end{pmatrix}$

6.6.3   Inner (dot) Product of Tensors

An inner or dot product of two tensors is where we sum the components of a pair of tensors with the same index when they are multiplied together. This was introduced earlier in section 2.2.3 on page 10. *The operation reduces the rank of the product.* The nomenclature is given below, where A and B are vectors and C is a scalar. Notice in Cartesian tensor notation form on the RHS we have 0 free indices,

Property	Vector Notation	Cartesian Tensor Notation
Inner (dot) product (of two vectors to produce a scalar result)	$C = \vec{A} \cdot \vec{B}$	$C = A_i B_i$

Table 6.9: Inner Dot Product of Two Tensors

implying that the LHS must be of zero rank, and one repeated index, implying we have to apply our inner (dot) product operation.

Example 10

Two vectors are defined $A_i = (1 \quad 2 \quad 3)$ and $B_i = (3 \quad 2 \quad 1)$, the dot product is written by $C = A_i B_i$. At the component level then $C = A_1 B_1 + A_2 B_2 + A_3 B_3 = 10$.

Example 11

If you dot a vector with itself and take the square root you obtain the vector magnitude, e.g. $|A_i| = (A_i A_i)^{1/2} = (A_1 A_1 + A_2 A_2 + A_3 A_3)^{1/2}$.

Example 12

$U_i \frac{\partial T}{\partial x_i}$, $\frac{\partial U_i}{\partial x_i}$, $U_i U_i$, $\sigma_{ij}\sigma_{ij}$ and $\frac{\partial U_i}{\partial x_i}\frac{\partial U_i}{\partial x_i}$ are all examples of dot product operations that produce a scalar quantity. Looking at each of these expressions, there are no free indices, therefore each of these quantities *must* be a scalar.

Example 13

We can "dot" tensors of different types, to produce a tensor a rank equal to the highest tensor in the expression, less the number of dot operations, for instance $A_i = B_j C_{ij}$. Here we can see the tensor equation is valid (same free index in each term) but the repeated j index on the RHS implies a dot product operation. In component form this is defined

$$A_i = \begin{pmatrix} A_1 & A_2 & A_3 \end{pmatrix} = B_j C_{ij} = \begin{pmatrix} B_1 \\ B_2 \\ B_3 \end{pmatrix} \begin{pmatrix} C_{11} & C_{12} & C_{13} \\ C_{21} & C_{22} & C_{23} \\ C_{31} & C_{32} & C_{33} \end{pmatrix} =$$

$$\begin{pmatrix} (B_1 C_{11} + B_2 C_{12} + B_3 C_{13}) & (B_1 C_{21} + B_2 C_{22} + B_3 C_{23}) & (B_1 C_{31} + B_2 C_{32} + B_3 C_{33}) \end{pmatrix}$$

Note how all the $j = 1$, $j = 2$ and $j = 3$ elements of B and C are grouped and eliminated, leaving only the i components, which are the three elements of the row vector A_i.

6.6.4 Cross Product of *Vectors*

All of the operations listed in section 6.6 thus far apply equally well to tensors of arbitrary order, however the cross product only applies to rank 1 tensors, or, more traditionally, vectors. It is also considerably easier to visualise geometrically and understand in vector notation. Geometrically, imagine two displacement vectors emerging from a common origin. The two vectors will define a plane. The cross product of two vectors defines a third vector normal to that plane. In other words it is normal to both of these vectors. In thermofluids, as part of the curl operator (section 6.7.3, below), it defines the vorticity (local rotation rate) of the velocity field. The Cartesian tensor equivalent operation does not have an obvious physical interpretation and makes use of the alternating unit tensor (section 6.4.4.2 on page 96). The operation relies on the concept of 'handedness' and traditionally 'right handed' convention is followed. This defines the positive direction of rotation. So if vector $\vec{A}$ is the first finger of your right hand and vector $\vec{B}$ is your second finger, then the cross product $\vec{C} = \vec{A} \times \vec{B}$ is a vector in the direction of your thumb. A left handed coordinate system would point $\vec{C}$ in the opposite direction. Note also this operation is not commutative, ie $\vec{A} \times \vec{B} \neq \vec{B} \times \vec{A}$, rather $\vec{A} \times \vec{B} = -\vec{B} \times \vec{A}$.

Property	Vector Notation	Cartesian Tensor Notation
Cross Product	$\vec{C} = \vec{A} \times \vec{B}$	$C_i = \epsilon_{ijk} A_j B_k$

Table 6.10: *Cross Product of two Vectors*

Example 14

$$\vec{A} = \begin{pmatrix} A_1 & A_2 & A_3 \end{pmatrix} = \vec{B} \times \vec{C} = \begin{pmatrix} B_2 C_3 - B_2 C_3 \\ B_3 C_1 - B_1 C_3 \\ B_1 C_2 - B_2 C_1 \end{pmatrix}$$

6.7 Cartesian Tensor Field Operations

Up till now, our tensor operations can operate on a point tensor or field tensor. Now we introduce a few operations that only operate on tensor fields. Field operations define how information changes spatially. The rate of change in one direction is a Gradient. The spreading from a point or line is a Divergence. The amount of rotation about a point is the Curl, and so on. You have already met one, the gradient operator. Like the gradient operator, which is physically important because it defines the direction and magnitude of mass, energy and momentum diffusion fluxes, other field operators are very important in defining the fundamental equations of continuum mechanics. So, as with the general tensor operators, we list the most important, starting with the gradient operator for completeness.

6.7.1 Gradient Operator

The gradient of a tensor results in a tensor of one order larger. The gradient of tensor u_i is the derivative of each of the velocity components in the x, y ,z directions and generates the velocity gradient tensor.

Property	Vector Notation	Cartesian Tensor Notation
Gradient (of a vector)	$\vec{C} = \vec{\nabla}\vec{u}$	$C_{ji} = \frac{\partial}{\partial x_j} u_i$

Table 6.11: *Gradient Operator*

6.7.2 Divergence Operator

The divergence of the tensor u_i is the inner (dot) product of the gradient operator and u_i. Because of the repeated index, summation in implied, for instance $C = \frac{\partial}{\partial x_i} u_i = \frac{\partial}{\partial x_1} u_1 + \frac{\partial}{\partial x_2} u_2 + \frac{\partial}{\partial x_3} u_3$. The divergence operator is very important because it is the most important term in the conservation of mass equation.

Property	Vector Notation	Cartesian Tensor Notation
Divergence (of a vector)	$C = \vec{\nabla} \cdot \vec{u}$	$C = \frac{\partial}{\partial x_i} u_i$

Table 6.12: *Divergence Operator*

6.7.3 Curl Operator

The curl operator define the local rotation about a point and like the cross product, it is better thought of using vector notation. In thermofluids it relates the vorticity field to the velocity field of a fluid flow. The vorticity can be thought of as a vector having magnitude equal to the maximum "circulation" at each point and to be oriented perpendicularly to this plane of circulation for each point.

Property	Vector Notation	Cartesian Tensor Notation
Curl (of a vector)	$\vec{\omega} = \vec{\nabla} \times \vec{u}$	$\omega_i = \epsilon_{ijk} \frac{\partial u_k}{\partial x_j}$

Table 6.13: *Curl Operator*

6.7.4 Laplacian Operator

The Laplacian is a scalar operator, and is composed of the gradient operator dotted with itself.

Property	Vector Notation	Cartesian Tensor Notation
Laplacian (of a vector)	$C = \nabla^2 T$	$C = \dfrac{\partial^2}{\partial x_i^2} T$

Table 6.14: *Laplacian Operator*

6.7.5 Taylor Series

As noted in section 2.2.6 on page 12 a Taylor series is a method to define change of information from one position (or time) to another. A scalar field in 1D is interpolated

$$\phi|_x = \phi|_0 + \left.\frac{d\phi}{dx}\right|_0 dx + \frac{1}{2}\left.\frac{d^2\phi}{dx^2}\right|_0 dx^2 \ldots$$

A vector field in 1D as

$$\phi_1|_x = \phi_1|_0 + \left.\frac{d\phi_1}{dx}\right|_0 dx \ldots, \quad \text{or} \quad \phi_i|_x = \phi_i|_0 + \left.\frac{d\phi_i}{dx}\right|_0 dx \ldots$$

Similarly a vector field and 3D gives

$$\phi_1|_x = \phi_1|_0 + \left.\frac{\partial\phi_1}{\partial x_1}\right|_0 dx_1 + \left.\frac{\partial\phi_1}{\partial x_2}\right|_0 dx_2 + \left.\frac{\partial\phi_1}{\partial x_3}\right|_0 dx_3 \ldots, \quad \text{or} \quad \phi_i|_x = \phi_i|_0 + \left.\frac{\partial\phi_i}{\partial x_j}\right|_0 dx_j \ldots$$

6.8 Properties of Cartesian 2nd order tensors

Say we have some tensor,

$$b_{ij} = \begin{pmatrix} b_{11} & b_{12} & b_{13} \\ b_{21} & b_{22} & b_{13} \\ b_{31} & b_{32} & b_{33} \end{pmatrix} = \begin{pmatrix} 1 & 2 & 3 \\ 4 & 5 & 6 \\ 7 & 8 & 9 \end{pmatrix}.$$

First we can decompose this into an isotropic part b_{ij}^I and a deviatoric part b_{ij}'. The isotropic part of b_{ij} is a third of the trace of b_{ij} multiplied by the kronecker delta.

$$b_{ij}^I = \frac{1}{3} b_{kk}\delta_{ij} = \begin{pmatrix} b_{kk}/3 & 0 & 0 \\ 0 & b_{kk}/3 & 0 \\ 0 & 0 & b_{kk}/3 \end{pmatrix} = \frac{15}{3}\begin{pmatrix} 1 & 0 & 0 \\ 0 & 1 & 0 \\ 0 & 0 & 1 \end{pmatrix} = \begin{pmatrix} 5 & 0 & 0 \\ 0 & 5 & 0 \\ 0 & 0 & 5 \end{pmatrix}.$$

The deviatoric part of b_{ij} is the difference between b_{ij} and b_{ij}^I

$$b_{ij}' = b_{ij} - b_{ij}^I = \begin{pmatrix} b_{11} - b_{kk}/3 & b_{12} & b_{13} \\ b_{21} & b_{22} - b_{kk}/3 & b_{23} \\ b_{31} & b_{32} & b_{33} - b_{kk}/3 \end{pmatrix} = \begin{pmatrix} -4 & 2 & 3 \\ 4 & 0 & 6 \\ 7 & 8 & 4 \end{pmatrix}.$$

By definition the deviatoric part of the tensor has zero trace: $b_{kk}' = 0$. Note however this does NOT mean the individual diagonal components are zero. The deviatoric part may be further decomposed into two further tensor types. A 'symmetric' deviatoric tensor:

$$b_{ij}^S = \frac{1}{2}\left(b_{ij}' + b_{ji}'\right) = b_{ji}^S = \begin{pmatrix} b_{11}^S & b_{12}^S & b_{13}^S \\ b_{21}^S = b_{12}^S & b_{22}^S & b_{23}^S \\ b_{31}^S = b_{13}^S & b_{32}^S = b_{23}^S & b_{33}^S \end{pmatrix} = \begin{pmatrix} -4 & 3 & 5 \\ 3 & 0 & 7 \\ 5 & 7 & 4 \end{pmatrix}.$$

And an 'anti-symmetric' deviatoric tensor:

$$b_{ij}^A = \frac{1}{2}\left(b_{ij}' - b_{ji}'\right) = -b_{ji}^A = \begin{pmatrix} 0 & b_{12}^A & b_{13}^A \\ b_{21}^A = -b_{12}^A & 0 & b_{23}^A \\ b_{31}^A = -b_{13}^A & b_{32}^A = -b_{23}^A & 0 \end{pmatrix} = \begin{pmatrix} 0 & -1 & -2 \\ 1 & 0 & -1 \\ 2 & 1 & 0 \end{pmatrix}.$$

Therefore, a 2 rank tensor may be decomposed to isotropic, symmetric and anti-symmetric 2 rank tensors, e.g.

$$b_{ij} = b_{ij}^I + b_{ij}^S + b_{ij}^A = \underbrace{\frac{1}{3}b_{kk}\delta_{ij}}_{isotropic} + \underbrace{b_{ij}^S}_{symmetric} + \underbrace{b_{ij}^A}_{anti-symmetric}$$

As we find out out in chapter 8, the velocity gradient tensor (ie b_{ij}) can be decomposed into the isotropic, symmetric and anti-symmetric tensors that have direct correspondence with physical processes.

6.9   General Orthogonal Coordinate Systems

In chapter 8 on page 111 the conservation equations of mass, momentum and energy are defined for a Cartesian coordinate system and also Cartesian tensor Notation, because these are the most common and also the most easy to use. In some cases however other coordinate systems are required, the most common being cylindrical. First transformations from Cartesian to cylindrical and spherical systems are defined.

6.9.1   Vector Identities
Here $\vec{A}$ is a vector and ϕ is a scalar. The following identities always hold.

$$\nabla \cdot \left(\phi\vec{A}\right) \equiv \phi\left(\nabla \cdot \vec{A}\right) + \vec{A} \cdot (\nabla\phi) \tag{6.2}$$

$$\nabla \times \left(\phi\vec{A}\right) \equiv \phi\left(\nabla \times \vec{A}\right) - \vec{A} \times (\nabla\phi) \tag{6.3}$$

$$\nabla \times (\nabla\phi) \equiv 0 \tag{6.4}$$

$$\nabla \cdot (\nabla \times \phi) \equiv 0 \tag{6.5}$$

6.9.2   Cartesian to Cylindrical Coordinates
Cartesian (x,y,z) to cylindrical (r,θ,z) coordinates are related by

$$x = r\cos\theta, y = r\sin\theta, z = z \tag{6.6}$$

$$r = (x^2 + y^2)^{1/2}, \theta = \arctan(y/x), z = z \tag{6.7}$$

Cartesian (x,y,z) to cylindrical (r,θ,z) velocity components by

$$u = u_r\cos\theta - u_\theta\sin\theta, v = u_r\sin\theta + u_\theta\cos\theta, w = w \tag{6.8}$$

$$u_r = u\cos\theta + v\cos\theta, u_\theta = -u\sin\theta + v\cos\theta, w = w \tag{6.9}$$

and the axis vectors defining the coordinate directions by

$$\vec{e_r} = \cos\theta\vec{e_x} + \sin\theta\vec{e_y} \tag{6.10}$$

$$\vec{e_\theta} = -r\sin\theta\vec{e_x} + r\cos\theta\vec{e_y} \tag{6.11}$$

$$\vec{e_z} = \vec{e_z} \tag{6.12}$$

6.9.3 Cylindrical Operators

Gradient Operator

$$\nabla\phi = \frac{\partial\phi}{\partial r}\hat{e_r} + \frac{1}{r}\frac{\partial\phi}{\partial\theta}\hat{e_\theta} + \frac{\partial\phi}{\partial z}\hat{e_z} \tag{6.13}$$

Divergence Operator

$$\nabla\cdot\vec{A} = \frac{1}{r}\frac{\partial}{\partial r}(rA_r) + \frac{1}{r}\frac{\partial A_\theta}{\partial\theta} + \frac{\partial A_z}{\partial z} \tag{6.14}$$

Curl Operator

$$\nabla\times\vec{A} = \left(\frac{1}{r}\frac{\partial A_z}{\partial\theta} - \frac{\partial A_\theta}{\partial z}\right)\hat{e_r} + \left(\frac{\partial A_r}{\partial z} - \frac{\partial A_z}{\partial r}\right)\hat{e_\theta} + \frac{1}{r}\left(\frac{\partial}{\partial r}(rA_\theta) - \frac{\partial A_r}{\partial\theta}\right)\hat{e_z} \tag{6.15}$$

Laplace Operator

$$(\nabla\cdot\nabla)\phi = \frac{1}{r}\frac{\partial}{\partial r}\left(r\frac{\partial\phi}{\partial r}\right) + \frac{1}{r^2}\frac{\partial^2\phi}{\partial\theta^2} + \frac{\partial^2\phi}{\partial z^2} \tag{6.16}$$

6.9.4 Cartesian to Spherical Coordinates

Cartesian (x,y,z) to spherical (r,θ,ϕ) coordinates are related by

$$x = r\sin\theta\cos\phi \tag{6.17}$$

$$y = r\sin\theta\sin\phi \tag{6.18}$$

$$z = r\cos\theta \tag{6.19}$$

and the unit vectors defining the coordinate directions by

$$\vec{e_r} = \sin\theta\cos\phi\vec{e_x} + \sin\theta\sin\phi\vec{e_y} + \cos\theta\vec{e_z} \tag{6.20}$$

$$\vec{e_\theta} = r\cos\theta\cos\phi\vec{e_x} + r\cos\theta\sin\phi\vec{e_y} - r\sin\theta\vec{e_z} \tag{6.21}$$

$$\vec{e_\phi} = -r\sin\theta\sin\phi\vec{e_x} + r\sin\theta\cos\phi\vec{e_y} \tag{6.22}$$

6.10 Summary and Further Reading

A brief outline has been provided, further details are available in the book by Rutherford Aris. The text by Pope is also excellent. Further operators in spherical, general orthogonal coordinate systems, complete sets of conservation equations and stress tensors for these systems are given in Bird, Stewart and Lightfoot. Note they define the viscous stress tensors with a negative sign, and use a negative sign in the conservation equations to match what is presented in section 8.8 on page 115. The focus of this book remains on Cartesian coordinate systems, because it is the easiest way to discuss the physics. The reader should note however additional forces appear when for instance the momentum equations are recast into cylindrical coordinate systems. From your experience on fairground rides, these additional forces are real, not simply mathematical leftovers, and in the case of momentum conservation arise due to the curvature of the coordinate system.

7 | Flow Deformation and Rotation

7.1   Introduction

In section 2.6.3 on page 24 the concept of relating a force due to deformation (shearing) of a fluid was made in one dimension. Equation 2.13 made the assumption that the force-deformation relationship was linearly proportional, and the proportionality constant the fluid dynamic viscosity, μ. This linearity defines the fluid as *Newtonian*. Here the concept of deformation is generalised and also to three dimensions and the stress is defined in it's proper tensorial form, ready for inclusion in general conservation equations defined in chapter 8 on page 111. The concept of local and global fluid rotation is also discussed, which needs to understood before defining potential flow in chapter 9 on page 117.

Figure 7.1 shows ways in which a fluid element can move (translate and rotate) and deform (strain). Usually, these all happen at once. Three types of strain are possible. Shear (introduced in section 2.6.3 page 24) and Plane Strain are volume conserving and are appropriate for incompressible fluids. In addition compression/expansion is considered, and this is an important link between the thermodynamics content when we considered the first law of thermodynamics (section 2.3.1, page 15) and the control volume based analysis of energy conservation *rate* (section 4.7, page 63).

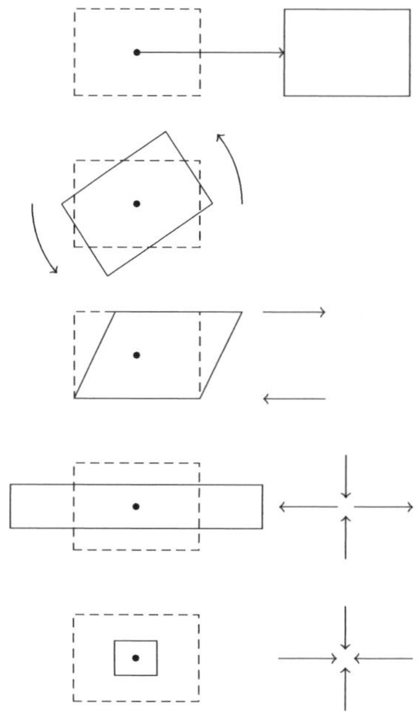

Figure 7.1: *Translation, Rotation and Deformation of a Fluid Element*

If we take two points in an *incompressible* fluid, they might be two corners of one of the systems shown in Fig. 7.1, and define point Q, in terms of displacement from point P, such that $x_i(Q) = x_i(P) + \delta x_i$, $u_i(Q) = u_i(P) + \delta u_i$ etc. The velocities at P and Q may be related by

$$u_i(x_j + \delta x_j) = u_i(x_j) + \left.\frac{\partial u_i}{\partial x_j}\right|_P \delta x_j \ldots \tag{7.1}$$

which we note is a Taylor series. Recall the velocity gradient tensor is defined

$$\frac{\partial u_i}{\partial x_j} = \begin{pmatrix} \partial u_1/\partial x_1 & \partial u_1/\partial x_2 & \partial u_1/\partial x_3 \\ \partial u_2/\partial x_1 & \partial u_2/\partial x_2 & \partial u_2/\partial x_3 \\ \partial u_3/\partial x_1 & \partial u_3/\partial x_2 & \partial u_3/\partial x_3 \end{pmatrix}$$

Where the gradient of each velocity *component* is a vector, for instance for the u_1 component $\partial u_1/\partial x_j =$

$(\partial u_1 / \partial x_1 \quad \partial u_1 / \partial x_2 \quad \partial u_1 / \partial x_3)$. The velocity at Q defined by equation 7.1 may be modified,

$$u_i\left(x_j + \delta x_j\right) = u_i\left(x_j\right) + \frac{\partial u_i}{\partial x_j}\partial x_j = \underbrace{u_i\left(x_j\right)}_{(a)} + \underbrace{\frac{1}{2}\left(\frac{\partial u_i}{\partial x_j} - \frac{\partial u_j}{\partial x_i}\right)\partial x_j}_{(b)} + \underbrace{\frac{1}{2}\left(\frac{\partial u_i}{\partial x_j} + \frac{\partial u_j}{\partial x_i}\right)\partial x_j}_{(c)} =$$

$$u_i\left(x_i\right) + \Omega_{ij}\partial x_j + S_{ij}\partial x_j.$$

Term (a) defines the translation of the point due to the velocity field. Term (b) is the rotation tensor Ω_{ij} and defines the angular velocity about the x_k axis at that point. Term (c) is the Strain Rate/Deformation rate tensor S_{ij} and includes linear (shear) strain and plane strain.

Referring to the properties of 2nd order tensors (section 6.8 page 101) we find that any 2nd order tensor (the velocity gradient for instance) can be decomposed into *three* parts. This reveals that bulk compression/expansion stress is also part of the velocity gradient tensor.

$$\frac{\partial u_i}{\partial x_j} = \underbrace{\frac{1}{3}\frac{\partial u_k}{\partial x_k}\delta_{ij}}_{isotropic} + \underbrace{S_{ij}}_{symmetric} + \underbrace{\Omega_{ij}}_{anti-symmetric}$$

This is the general form of the velocity gradient decomposition for a compressible fluid.

7.2   Linear and Plain Strain

Note previously in section 2.6.3, page 24 we only defined shear strain in a uni-directional flow. Here we define linear strain generally, in the 1-2 plane $S_{12} = \frac{1}{2}\left(\frac{\partial u_1}{\partial x_2} + \frac{\partial u_2}{\partial x_1}\right)$ and as shown by Fig. 7.2 the rate of strain of the velocity field is

$$\frac{1}{2}\left(\frac{\text{deformation in } x_1}{\text{length in } x_2} + \frac{\text{deformation in } x_2}{\text{length in } x_1}\right) \Big/ \text{time} = \frac{1}{2}\left(\frac{\frac{\partial u_1}{\partial x_2}dx_2 dt}{dx_2} + \frac{\frac{\partial u_2}{\partial x_1}dx_1 dt}{dx_1}\right) \Big/ dt =$$

$$\frac{1}{2}\left(\frac{\partial u_1}{\partial x_2} + \frac{\partial u_2}{\partial x_1}\right)$$

Fig. 7.3 shows the simpler plane strain case, and here

$$S_{11} = \left(\frac{\text{extension in } x_1}{\text{original length}}\right) \Big/ \text{time} = \frac{\frac{\partial u_1}{\partial x_1}dx_1 dt}{dx_1 dt} = \frac{\partial u_1}{\partial x_1}$$

7.3   Bulk Compression/Expansion Strain

The compression/expansion strain arises from the isotropic term of the 2nd order tensor decomposition, and is zero in incompressible fluids, by virtue of the continuity equation (equation 4.4 on page 63), since for pure incompressible liquids.

$$\frac{\partial u_k}{\partial x_k} = \frac{\partial u_1}{\partial x_1} + \frac{\partial u_2}{\partial x_2} + \frac{\partial u_3}{\partial x_3} = \frac{\partial u}{\partial x} + \frac{\partial v}{\partial y} + \frac{\partial w}{\partial z} = 0 \tag{7.2}$$

Therefore, equation 7.2 defines mass conservation for an incompressible fluid.

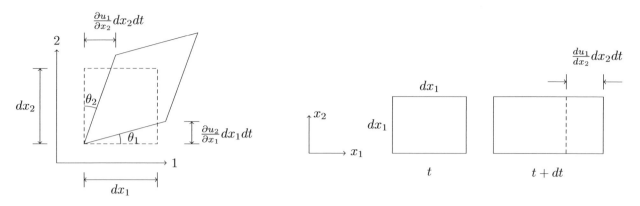

Figure 7.2: *Two dimensional Shear Strain*

Figure 7.3: *Plane Strain*

7.4 The Stress Tensor

Here we simply relate the viscous stress tensor to the velocity gradient tensor using the dynamic viscosity, in other words a multidimensional form of equation 2.13 on page 25.

$$\tau_{ij} = 2\mu S_{ij} \tag{7.3}$$

The viscous stress tensor is the sum of the normal and shear stresses in the fluid and is denoted τ_{ij}, since the first index denotes the direction of the force and the second index the area normal of the face over which the force acts. For instance the total viscous force in the 1 direction is $f_1 = \tau_{1j}A_j = \tau_{11}A_1 + \tau_{12}A_2 + \tau_{13}A_3$, the sum of a normal and two shear stresses.

The total stress tensor σ_{ij} is defined such that the force required to oppose the pressure in the fluid and any stress due to plane strain. Considering the pressure component, when a fluid is at rest the normal stress must balance the hydrostatic pressure due to the weight of the fluid, thus $\sigma_{11} = \sigma_{22} = \sigma_{33} = -p$, $\sigma_{kk}/3 = -p$ and $\sigma_{ij} = -p\delta_{ij}$, therefore,

$$\sigma_{ij} = -p\delta_{ij} + \tau_{ij} = -p\delta_{ij} + 2\mu S_{ij} + \lambda \frac{\partial u_k}{\partial x_k}\delta_{ij}.$$

Since we know $\sigma_{kk}/3 = -p$, $\lambda = -(2/3)\mu$ and this gives our complete form for the total stress tensor.

$$\sigma_{ij} = -\left(p + \frac{2}{3}\mu S_{kk}\right)\delta_{ij} + \tau_{ij} \tag{7.4}$$

This immediately allows us to work out the total force on the surface of a control volume by $f_i = \sigma_{ij}A_j$. This is composed of normal total stress which is the force per unit area due to the pressure, the compression/expansion rate and the rate of linear strain normal to the surface.

$$\sigma_{11} = -p - \frac{2}{3}\mu\frac{\partial u_k}{\partial x_k} + 2\mu\frac{\partial u_1}{\partial x_1}.$$

Likewise the shear total stress is composed to the strain rate in the two direction parallel to the surface.

$$\sigma_{12} = \mu\left(\frac{\partial u_1}{\partial x_2} + \frac{\partial u_2}{\partial x_1}\right)$$

7.5   The Vorticity Vector and the Rotation Tensor

Because the rotation tensor Ω_{ij}, as a measure of the angular velocity about the k axis it is not at all intuitive, a 'vorticity' vector is also defined, ω_k. By convection its magnitude is twice the rotation rate. Vorticity is a measure of the degree of local rotation in the fluid and has dimensions $[T]^{-1}$. It is defined in vector form as $\vec{\omega} = \nabla \times \vec{u}$ and in tensor notation,

$$\omega_i = \epsilon_{ijk} \frac{\partial u_k}{\partial x_j} = \left(\frac{\partial u_3}{\partial x_2} - \frac{\partial u_2}{\partial x_3} \quad \frac{\partial u_1}{\partial x_3} - \frac{\partial u_3}{\partial x_1} \quad \frac{\partial u_2}{\partial x_1} - \frac{\partial u_1}{\partial x_2} \right) = (-2\Omega_{23} \quad -2\Omega_{13} \quad -2\Omega_{12}). \quad (7.5)$$

In a 2D flow this vector is always normal to the flow plane $\omega_3 = \partial u_2/\partial x - \partial u_1/\partial y$. The relation between the vorticity vector and Cartesian coordinate directions is explained by reference to Fig. 7.4. By trigonometry $u_1 = -\omega_3 R sin\theta$ and $u_2 = \omega_3 R cos\theta$, $cos\theta = x_1/R$ and $sin\theta = x_2/R$ and so $u_1 = -\omega_3 x_2$ and $\partial u_1/\partial x_2 = -\omega_3$.

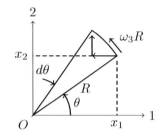

Figure 7.4: *Relating vorticity to the rotation tensor*

7.6   Circulation

Circulation is a macroscopic measure of fluid rotation, whereas the vorticity is the rotation at a point. It is used primarily in the prediction of lift using the Kutta-Jowkowsky Theorem defined in equation 9.8 on page 123 solving a potential flow as discussed in chapter 9. The circulation is simply sum of the dot product of the velocity and a line integral, where the lines (displacement vectors) form a closed curve.
Since the shape enclosed by the curve can be any shape and area, the amount of circulation can vary. It is defined as

$$\Gamma = -\int \vec{u} \cdot d\vec{x} = -\int (\nabla \times \vec{u}) \cdot d\vec{A} \quad (7.6)$$

and some texts define positive circulation in the opposite sense. As noted by equation 7.6 the circulation can also be defined in terms of the vorticity *in* the area at a point dotted with the vectorial area element of that point. This leads immediately to the statement that if the flow has no vorticity inside any closed curve, then the circulation is also zero. In another sense, the vorticity can be thought of as circulation per unit area. This has a caveat, that the enclosed area must be infinitely reducible to a point, and is illuminated in the discussion of the vorticity and circulation in a free vortex, section 7.8, below. The point that the circulation is used to generate lift from a potential flow solution of an irrotational flow (chapter 9) should be fully understood!

7.7   Solenoidal and Irrotational Flows

Often the velocity field can be characterised by precise mathematical definitions that have both physical meaning and permit simplification and in some cases analytical solution of the flow field, as in chapter 9 on page 117. Fluid flows are said to be 'solenoidal' or 'divergence free' when equation 7.2 is true. Flows of all pure incompressible fluids are solenoidal. Another condition is 'irrotational' or 'curl free' flow, where the fluid flow field has no rotation, in other words equation 7.5 is set to zero. Viscous dominated flows are typically highly rotational, but in many cases, flows away from walls have little if any rotation.

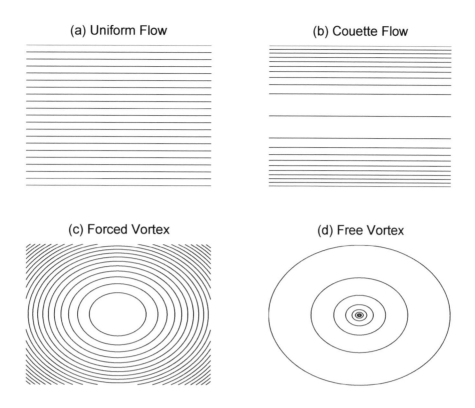

Figure 7.5: *Streamline for Steady Uniform, Couette, Forced and Free Vortex Flows*

7.8 Flow Examples to Demonstrate Deformation, Vorticity and Circulation

Here four examples illuminate perhaps unexpected results, and each are illustrated in Figure 7.5. All are 2D and may be defined in $x - y$ or $r - \theta$ coordinates. It is good practice to define the velocity field in both and then do all the further calculations in both too, this self checks everything. For reference the continuity and vorticity for these are as follows :

Property	$x - y$	$r - \theta$
Continuity	$\frac{\partial u}{\partial x} + \frac{\partial v}{\partial y} = 0$	$u_r + r\frac{\partial u_r}{\partial r} + \frac{\partial u_\theta}{\partial \theta} = 0$
Vorticity	$\frac{\partial v}{\partial x} - \frac{\partial u}{\partial y} = \omega$	$\frac{u_\theta}{r} + \frac{\partial u_\theta}{\partial r} - \frac{1}{r}\frac{\partial u_r}{\partial \theta} = \omega$

Table 7.1: *Solenoidal and Irrotational Conditions for 2D $x - y$ and $r - \theta$ Coordinate Systems*

7.8.1 Example 1 : Steady Uniform Flow

As shown in Figure 7.5a the velocity field is $u_i = [u_o, 0, 0]$ and hence the flow is both solenoidal and irrotational by virtue of equations 7.2 and 7.5 respectively. Also because it is irrotational, the circulation is also zero.

7.8.2 Example 2 : Steady Couette Flow (section 5.2 on page 76)

As shown in Figure 7.5b the velocity field could be represented by $u_i = [u_o y, 0, 0]$, the flow velocity is parallel but linearly related to y. We can see that the flow is solenoidal but it is not irrotational, since $\omega_i = [0, 0, -u_o]$. So even though the fluid flow is parallel, it is *locally* rotating. We can also see the circulation is non-zero, since the velocities on the two horizontal parts of the curve are unequal.

7.8.3   Example 3 : Forced Vortex (solid body rotation)

Here we analyse the fluid rotating about a cylindrical axis, such that the velocity field is rotating at constant angular speed $\dot{\theta}$ and the angular velocity is defined by $u_\theta = \dot{\theta}r$ and $u_z = u_r = 0$. We find that this flow field is solenoidal but rotational. The key to understanding this is that here the fluid is globally and locally rotating. If we replaced the fluid with a steel bar we would get the same result. Fluid elements nearby to a central fluid element are not moving relative to that central element. The vorticity at a point is the *local* rotation or spin of the fluid element at that point and, from equation 7.5 is simply twice the angular velocity, therefore $\omega_z = u_\theta \hat{\theta}$ and the circulation is $\Gamma = -2\pi r u_\theta$, where $2\pi r$ is the length of the curve enclosing the circulating fluid.

7.8.4   Example 4 : Free Vortex

We now analyse another rotation about an axis but in this case the angular velocity decays with distance from the axis such that $u_\theta = \dot{\theta}/r$ and $u_z = u_r = 0$. We find that this flowfield is solenoidal and irrotational, despite each fluid element rotating about a central point. By virtue of equation 7.6, because the flow is irrotational, the vorticity vector everywhere is also zero, and the circulation *should* be zero. But it isn't, if we work it out it is $\Gamma = -2\pi\dot{\theta}$. How can this be ? An additional caveat to the definition of circulation is that the area enclosed must be infinitely reducible to a point and the flow properties at that point must be well behaved. For the free vortex, as $r \to 0$ then $u_\theta \to \infty$, thus our circulation measure for a free vortex is invalid.

7.8.5   Discussion

As discussed above, in examples 2 and 3 the local fluid element is rotating, locally, while it is being translated. In example 2 it is being translated rectilinearly, while in example 3 it is being transported in a circular motion. In examples 1 and 4 it is just being translated. It is important to distinguish between rotation of fluid along a streamline (vorticity) and the curvature of the streamline itself. Hopefully this shows that it is possible for parallel flows to be (locally) rotating, and rotational flows to be (locally) irrotational. Another way to consider the irrotational condition is whether viscosity is present. Examples 2 and 3 are examples of real life flows, where viscous forces are important. In general viscous flows have local rotation. So, for example considering the flow over an aircraft wing, only the viscous dominated boundary layer region is rotational.

7.9   Summary and Further Reading

A brief introduction to the subject of 'kinematics' is provided. Explanation of the strain terms is needed to understand stress in 3D in preparation for derivation of the conservation equations in chapter 8 on page 111. Explanation of the rotational terms and circulation is required to understand potential flow in chapter 9 on page 117. Further reading is available in many of the older fluids texts, the text by Currie is recommended.

8 | Conservation Equations

8.1 Introduction

In this chapter we make use of the power of tensor notation to derive the full conservation laws of mass, momentum and energy over a 3D Eulerian control volume succinctly and precisely. To explain the word "full" consider the following.

1. In section 4.8 on page 66 we defined the conservation of momentum assuming the fluid is inviscid and steady, and therefore this equation related the inertial momentum change to the force applied to it.
2. In section 4.9.1 on page 68 we related the weight of the fluid to the hydrostatic pressure felt by the fluid, assuming the fluid was at rest.
3. In section 5.2 on page 76 we applied a force balance to a control volume containing a viscous fluid in a Couette flow and showed that the only force present was due to a shear stress.
4. In section 5.4 on page 77 we applied a force balance to a control volume containing a viscous fluid in a Poisuelle flow, and showed a pressure force balanced the viscous force.

All of these cases noted above are special cases of a general momentum conservation equation, which combines them all. The special cases have been introduced first because each one of them introduces you to a distinct physical process that requires understanding at the conceptual level. In a similar way energy was introduced for a system (fixed mass) in section 2.3.1, page 15 and later, for a control volume the steady flow energy equation (section 4.7 page 63) assumed the fluid is inviscid and steady. We touched on viscous losses in the mechanical energy equation, but did not really explain the mechanism.

8.2 Basis for Eulerian Conservation Equations

We wish to convert our fundamental Lagrangian conservation equations for (extensive) mass, momentum and energy for a (system) fixed mass into Eulerian (intensive) density, velocity and specific energy conservations for a fixed (control) volume. Lagrangian and Eulerian coordinate systems are defined in section 2.2.10 on page 14. Systems and control volumes are defined near section 2.3.4 on page 16, extensive and intensive variables in section 2.3.3, page 15. Transferring information from a finite mass (a system) to a control volume, using the Reynolds Transport Theorem is defined in section 2.3.6 on page 17. We follow this basis below, but in subsequent sections we define the conservation of a general intensive property over a fixed (control) volume directly. All our fundamental conservation equations have the form

$$\frac{D\phi}{Dt} = f(\phi), \tag{8.1}$$

where ϕ represents mass, momentum or energy and $f(\phi)$ is a general source term that creates or destroys ϕ. For the momentum equation this might be acceleration due to gravity for Newton's 2nd Law and the motion of his apple. For the rate form of the 1st law of thermodynamics this would the energy and work transfers across the system surface. To convert equation 8.1 to a control volume basis we do two things:

1. We apply the Reynolds transport theorem to the LHS using equation 2.11 on page 18.
2. We convert our extensive variable ϕ to its intensive form φ using equation 2.10 on page 16.

This gives equation 8.2, a fundamental and general conservation equation for a control volume

$$\frac{\partial}{\partial t} \int \rho\varphi \, \partial V + \int \rho\varphi u_i \, \partial A_i = \int \rho f(\varphi) \, \partial V \qquad (8.2)$$

To define specific conservation equations for density, velocity and specific energy, or indeed any other conserved quantity we simply define what φ is and the relevant source terms $f(\varphi)$. Equation 8.2 conserves an intensive variable φ for a *finite* volume, and as such is the basis for the finite volume method, used in computational fluid dynamics, and discussed in chapter 14. Equation 8.2 is also commonly described as an *integral* conservation equation for obvious reasons. The differential form, defined below for specific variables is the more common form in textbooks, and is obtained by taking the limit of $\delta V \to 0$. In that case the resulting *differential form* is defined.

$$\frac{\partial \rho\varphi}{\partial t} + \frac{\partial \rho\varphi u_i}{\partial x_i} = \rho f(\varphi) \qquad (8.3)$$

It is worth recalling at this point we have already derived convective and diffusive fluxes over a control volume for steady 1-D flow in section 2.6.5, page 27. In what follows, we have already defined the net convective flux over our control volume by virtue of our application of the Reynolds Transport Theorem. The next few sections proceeds as follows. First we define the conservation of mass, and simplifications, then we define a general diffusive flux, a component of all variables except density. Finally we then define the source terms particular to the velocity and specific energy variables that properly conserve these variables.

8.3 Conservation of Mass (or Continuity) Equation

The differential form follows directly from equation 8.2 by setting $\varphi = 1$, defining a general mass source/sink, and taking the limit as the volume goes to zero. In tensor notation this has units of kg/s and is

$$\frac{\partial \rho}{\partial t} + \frac{\partial \rho u_i}{\partial x_i} = S_\rho \qquad (8.4)$$

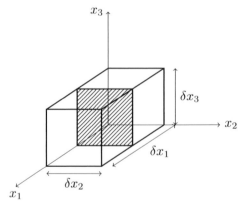

Figure 8.1: *A face at the centre of a control volume with a normal in the x_1 direction*

Typically $S_\rho = 0$ and if the fluid is incompressible the continuity equation simplifies to $\frac{\partial u_i}{\partial x_i} = 0$, the condition for solenoidal flow, as noted previously by equation 7.2 on page 106.

8.4 Conservation of Diffusion Flux over a Control Volume Surface for a Scalar

This has already been discussed in 1-D in section 2.6.5 page 27 and here we define the general 3D case for the general intensive variable φ. Fig. 8.1 shows a 3D control volume aligned with the axes of a Cartesian coordinate system and in it a *y-z* planar section of the control volume at $dx_1/2$. The diffusion flux needs to be conserved over the 2 faces in each of the 3 coordinate directions and is a general tensor form of the 1-D conservation of diffusion flux (over 2 faces) outlined earlier in section 2.6.5.3 on page 28. Taking thermal energy as our example the heat flow would be $\dot{q}$ in Js^{-1}, the heat *flux* in $Js^{-1}m^{-2}$, in the 1-direction is $\dot{q}_1 = -k\partial T/\partial x_1$. The heat *flow* Js^{-1}, at the control volume centre, in the 1-direction is $A\dot{q}_1 = \delta x_2 \delta x_3 q_1$. We extrapolate this information to the control volume faces using a Taylor series, and hence the diffusional heat *out flow* Js^{-1}, at $x + \delta x_1/2$, in the 1-direction is $-\delta x_2 \delta x_3 (\dot{q}_1 + (\delta x_1/2)(\partial \dot{q}_1/\partial x_1))$ and similarly the heat diffusion *in flow* Js^{-1}, at $x - \delta x_1/2$, in the 1-direction is $+\delta x_2 \delta x_3 (\dot{q}_1 - (\delta x_1/2)(\partial \dot{q}_1/\partial x_1))$. This gives the net heat diffusion *flow* Js^{-1}, in 1-direction is $+\delta x_1 \delta x_2 \delta x_3 ((\partial \dot{q}_1/\partial x_1))$ and in 3D.

$$+ \delta x_1 \delta x_2 \delta x_3 \left(\partial \dot{q}_i / \partial x_i\right). \tag{8.5}$$

8.5 Conservation of Diffusion Flux over a Control Volume Surface for a Vector

Here we take the example of the velocity field, and thus we are defining a rate of specific momentum transfer due to viscosity. The total stress tensor which expresses the viscous and pressure forces acting on the surfaces of the control volume was discussed in section 7.4 on page 107. Taking Fig. 8.1 as our exemplar, one normal and two shear forces (N m^{-2}, momentum flow per unit area), acting in the 1-direction at the CV centre are present, $-\sigma_{11}, -\sigma_{12}, -\sigma_{13}$. Using the same method as above the total force (N) over the six control volume faces, acting in the one direction is.

$$\delta x_1 \delta x_2 \delta x_3 \frac{\partial \sigma_{1j}}{\partial x_j} \tag{8.6}$$

8.6 Conservation of Momentum Equation

Here we define the general equation for momentum conservation for a Newtonian fluid, and are named the Navier-Stokes equations, for the three velocity components. These equations arise directly from Newton's Second Law (equation 2.8 page 15) for a system. These are expressed for a fixed volume (equation 8.3) and are generalized to include surface (viscous and pressure) forces as defined in section 8.5 and volumetric (here gravitational) forces, terms that comprise the RHS of equation 8.3. In this case the Navier-Stokes equations becomes

$$\frac{\partial}{\partial t}(\rho u_i) + \frac{\partial}{\partial x_j}(\rho u_i u_j) = -\frac{\partial}{\partial x_i}\left(p + \frac{2}{3}\mu \frac{\partial u_k}{\partial x_k}\right) + \frac{\partial \tau_{ij}}{\partial x_j} + \rho g_i \tag{8.7}$$

Which can be simplified for an incompressible pure fluid to:

$$\frac{\partial u_i}{\partial t} + u_j \frac{\partial u_i}{\partial x_j} = -\frac{1}{\rho}\frac{\partial p}{\partial x_i} + \nu \frac{\partial^2 u_i}{\partial x_j^2} + g_i$$

It should be noted that this equation can be simplified to define earlier force balances, for instance hydrostatics (section 4.9.1 page 68, by setting $u_i = 0$) pressure-inertia (section 4.8 page 66, by setting $\nu = 0$) and viscous flow (chapter 5, page 75, by setting the inertial and pressure terms (2nd term of the LHS, 1st term of the RHS) to zero). So, here are several key points.

- All the basic conceptual stuff you learnt in chapters 2 to 5 applies here and nothing more. You learnt all the physics already.
- In this chapter you haven't learnt anything new really, just how to combine pressure, viscous and inertial forces into one 'master' equation and generalise it to 3D.
- This is it for momentum conservation. It does not get any more complicated than this. Some of the finest minds on the planet are still studying it.

8.7   Conservation of Thermal Energy Equation

There are many forms of energy, and therefore many forms of energy equation. Here we derive a thermal energy equation from the rate form of the first Law and subtract the kinetic energy components. The kinetic energy rate equation is obtained directly by taking the dot product of equation 8.7 with u_i,

$$\rho\frac{\partial (u_k u_k/2)}{\partial t} + \rho u_i\frac{\partial (u_k u_k/2)}{\partial x_i} = u_i\rho g_i + u_i\frac{\partial \sigma_{ij}}{\partial x_j} \tag{8.8}$$

The rate form of the first law for total (stagnation) internal energy for a system, equation 2.9, page 15 expressed for a fixed volume, equation 8.3 is

$$\frac{\partial \rho e_{u0}}{\partial t} + \frac{\partial \rho e_{u0} u_i}{\partial x_i} = k\frac{\partial^2 T}{\partial x_i^2} - \dot{w}$$

where we have defined the diffusive flux of thermal energy as the heat transfer across the control volume boundary. For the work term we are undertaking an energy rate balance over a CV, so we consider work rate, i.e. power. Recall work = force × distance and work rate = power = force × velocity. We have work done on the volume and on the surface of the volume. Considering force in the 1 direction only, we transform tensor to force quantities by using vector area, e.g. $f_1 = A_j\sigma_{1j}$. The force in the $i = 1$ direction, components = $f_1 = A_1\sigma_{11} + A_2\sigma_{12} + A_3\sigma_{13}$. Work *rate*, a scalar (of order zero) must be, $\dot{w} = f_i u_i = A_j\sigma_{ij}u_i$. $A_j\sigma_{ij}u_i$ has two repeated indices, and has 9 components, and is therefore a scalar. Applying this over the six control volume faces as in section 8.4 page 112 we arrive at

$$\frac{\partial (\rho e_{u0})}{\partial t} + \frac{\partial (u_k\rho e_{u0})}{\partial x_k} = k\frac{\partial^2 T}{\partial x_k^2} + \rho u_k g_k + \frac{\partial \sigma_{kl}u_k}{\partial x_l}$$

Subtracting the kinetic energy rate equation gives us a thermal energy conservation equation

$$\frac{\partial \rho e_u}{\partial t} + \frac{\partial (\rho u_j e_u)}{\partial x_j} = k\frac{\partial^2 T}{\partial x_k^2} + \sigma_{ij}\frac{\partial u_i}{\partial x_j} \tag{8.9}$$

So here we have a time dependent term and the convective term on the LHS, which working back through the Reynolds Transport theorem would just be DE_T/Dt. On the RHS we have the heat and the work transfers across the control volume boundary, namely $\dot{q} - \dot{w}$.

8.7.1   Thermal Energy Work Terms

The work term (last term of the RHS of equation 8.9) can be further split into reversible work transfer (for instance compressing a gas in a piston, heating it up, and the reverse), and irreversible work. This latter term is always positive, and always converts motion into thermal energy, due to friction.

The last term of equation 8.9 can be further simplified into a reversible part (temperature may go up or down, for instance a frictionless piston) and an irreversible part (the temperature always rises due to friction) which is responsible for entropy generation due to viscous losses.

$$\sigma_{ij}\frac{\partial u_i}{\partial x_j} = -p\frac{\partial u_k}{\partial x_k} + \mu\Phi. \tag{8.10}$$

It can be shown (for instance Bird, Stewart and Lightfoot) that the viscous dissipation term Φ is composed entirely of velocity gradients and is always positive. Thus for viscous fluids there is always irreversible frictional conversion of kinetic energy into heat and therefore production of entropy.

8.8 Non-dimensional Conservation Equations

By following the procedure outlined in section 2.6.5.5 on page 29 we can define a non-dimensional general equation from equations 8.4, 8.7 and 8.9.

$$\frac{\partial}{\partial t^*}\left(\rho^*\varphi^*\right) + \frac{\partial}{\partial x_j^*}\left(\rho^*\varphi^*u_j^*\right) = A_\varphi\frac{\partial}{\partial x_j^*}\left(\Gamma_\varphi\frac{\partial\varphi^*}{\partial x_j^*}\right) + S_\varphi^*$$

We can also define a transport equation for a 'passive scalar', a chemical species that does not react, in terms of the mass fraction, m_a. And for mass, momentum and energy and species equations the coefficients are:

	A_φ	Γ_φ	S_φ^*
Mass $(\varphi^* = 1)$	1	0	0 (usually)
Momentum $(\varphi_i^* = u_i^*)$	$\frac{1}{\mathrm{Re}_o}$	μ^*	Incompressible: $\frac{1}{\mathrm{Re}_o}\frac{\partial}{\partial x_j^*}\left(\mu^*\left\{\frac{\partial u_j^*}{\partial x_i^*}\right\}\right) - \frac{\partial p^*}{\partial x_i^*} + \mathrm{Fr}_o\rho^*g_i^*$ Compressible: $\frac{1}{\mathrm{Re}_o}\frac{\partial}{\partial x_j^*}\left(\mu^*\left\{\frac{\partial u_j^*}{\partial x_i^*}\right\} - \frac{2}{3}\mu^*\frac{\partial u_k^*}{\partial x_k^*}\delta_{ij}\right) - \mathrm{Eu}_o\frac{\partial p^*}{\partial x_i^*} + \mathrm{Fr}_o\rho^*g_i^*$
Energy $(\varphi^* = e_u^*)$	$\frac{1}{\mathrm{Re}_o\mathrm{Pr}_o}$	$\frac{\mu^*}{\mathrm{Pr}^*}$	Incompressible: $\frac{\mathrm{Br}_o}{\mathrm{Pr}_o}\left\{\frac{\partial p^*}{\partial t^*} + u_i^*\frac{\partial p^*}{\partial x_i^*}\right\} + \frac{\mathrm{Ek}_o}{\mathrm{Re}_o}\mu^*\Phi^*$ Compressible: $(\gamma_o - 1)\,\mathrm{Ma}_o^2\left\{\frac{\partial p^*}{\partial t^*} + u_i^*\frac{\partial p^*}{\partial x_i^*}\right\} + \frac{\mathrm{Ek}_o}{\mathrm{Re}_o}\mu^*\Phi^*$
Species $(\varphi^* = m_a^*)$	$\frac{1}{\mathrm{Re}_o\mathrm{Sc}_o}$	$\frac{\mu^*}{\mathrm{Sc}^*}$	0 (usually)

Table 8.1: *Terms in the General Non-dimensional Conservation Equation*

Two forms are required because pressure has different meaning in compressible and incompressible fluids. For compressible fluids the reference pressure is taken as p_0, whereas for incompressible fluids the reference pressure is defined from the Bernoulli equation, namely $p_0 = \rho_0 u_0^2$. This shows that only the pressure gradient (not the pressure) is important for flow in incompressible fluids. The various new non-dimensional numbers that drop out of this analysis are as follows

Presenting the governing equations non-dimensionally really defines what the non-dimensional numbers are physically. For instance when $Re << 1$ then the inertial time dependent terms, pressure and body forces are all negligible and the Navier-Stokes equations simplify to the Stokes Equations. The 1-D form of these was introduced in section 5.2 on page 76. Likewise when $Re \to \infty$, the viscous forces are negligible and the Navier-Stokes equations simplify to the Euler Equations. These are the basis of potential flow discussed in chapter 9 on page 117.

8.9 Summary

This chapter is the link between the introductory material of the preceding chapters, which introduced the basic physical principles of thermofluids and more specialist material from this point on. Computational fluid dynamics, introduced in chapter 14 solves the integral versions of these equations directly using numerical methods. Therefore in one sense you can now solve any thermofluids problem. The equations can also be simplified as outlined above to recover key physical relationships as outlined in

$\left[\dfrac{\rho_o u_o x_o}{\mu_o}\right]$	$\left[\dfrac{\text{convective flux}}{\text{viscous flux}}\right]$	Reynolds	Re_o
$\left[\dfrac{C_{po}\mu_o}{k_o}\right]$	$\left[\dfrac{\text{viscous flux}}{\text{thermal flux}}\right]$	Prandtl	Pr_o
$\left[\dfrac{\rho_o D_{abo}}{\mu_o}\right]$	$\left[\dfrac{\text{viscous flux}}{\text{mass flux}}\right]$	Schmidt	Sc_o
$\left[\dfrac{p_o}{\rho_o u_o^2}\right]$	$\left[\dfrac{\text{pressure force}}{\text{inertia force}}\right]$	(half of) Euler	Eu_o
$\left[\dfrac{g_o x_o}{u_o^2}\right]$	$\left[\dfrac{\text{body force}}{\text{inertia force}}\right]$	(square of inverse) Froude	Fr_o
$\left[\dfrac{u_o^2}{\gamma_o R_o T_o}\right]$	$\left[\dfrac{\text{flow speed}}{\text{wave speed}}\right]$	Mach	Ma_o
$\left[\dfrac{\mu_o u_o^2}{k_o T_o}\right]$	$\left[\dfrac{\text{viscous heating}}{\text{heat conduction}}\right]$	Brinkman	Br_o
$\left[\dfrac{u_o^2}{C_{po} T_o}\right]$	$\left[\dfrac{\text{kinetic energy}}{\text{heat capacity}}\right]$	Eckert	Ek_o

Table 8.2: *Key Non-dimensional Numbers in the Governing Equations*

the previous chapters of this book.

9 | Potential Flow

9.1 Introduction

Prior to the development of advanced experimental and computational methods, potential flow was used extensively by aerodynamicists for the prediction of lift from wing sections. It is still a very useful technique to obtain good approximations to flow patterns cheaply and quickly where pressure and inertial forces dominate.. Several assumptions are required;

- Viscous forces and mixing processes are absent, this requires our fluid flow to be away from walls and not turbulent.

- The flow must be solenoidal. As discussed in section 7.3 on page 106, equation 7.2 requires the fluid to be incompressible.

- The flow is irrotational, that the vorticity is zero everywhere, in other words $\vec{\omega} = \nabla \times \vec{u} = 0$.

Although not strictly required, here we further define the flow to be steady and also 2D, in both Cartesian (x, y) and cylindrical (r, θ) coordinate systems. With the fluid and flow specified with these restrictions, the streamline definition (section 4.1.2 on page 59, equation 4.1) is valid and mass, momentum and energy conservation are defined by the solenoidal condition, equation 7.2 on page 106, the Euler equations and the Bernoulli equation 4.3 on page 61.

The potential flow approach describes the flow field in terms of solutions to the Laplace equation, and is able to remove the pressure velocity coupling. Solutions to the Laplace equation are linear and thus solutions from simple flows may be added together to form solutions to more complicated problems. It must be remembered however that a key assumption is the fluid is inviscid, so while the pressure distribution may be calculated, leading to prediction of the lift and form drag (section 5.7.5.1 on page 85), the friction drag (due to viscous shear stresses) is absent.

In what follows first the stream and velocity potential functions are defined, which lead to Laplace equation solutions. Then four fundamental base flow configurations are defined and are superimposed in various ways. Finally a coordinate transformation is introduced which permits the prediction of lift on aerofoil shapes and the Kutta condition is introduced which imposes a reality constraint on the flow over these shapes.

9.2 Stream Function

A stream function defines the mass flow distribution of an entire flowfield defined by a set of streamlines, discussed previously in section 4.1.2 on page 59. Recall, as defined by equation 4.1, the velocity vector is always parallel to the streamline at a point. The mass flux per unit width (into the page) could be defined by $\dot{m} = \rho \int u \, dy - \int v \, dx$ between two streamlines. Since the fluid is incompressible the stream function is defined $\psi = \int u \, dy - \int v \, dx$, in essence the stream function is a scalar volume flux distribution. Due to the incompressible fluid assumption our flow is solenoidal (equation 7.2), this requires the following relationship between the velocity components and the stream function for Cartesian and cylindrical systems :

$$u = +\frac{\partial\psi}{\partial y}, v = -\frac{\partial\psi}{\partial x} \tag{9.1}$$

$$u_r = +\frac{1}{r}\frac{\partial\psi}{\partial\theta}, u_\theta = -\frac{\partial\psi}{\partial r} \tag{9.2}$$

Note this is not entirely universal, some texts reverse the flow direction of the streamlines and thus the signs in equations 9.1 and 9.2. The stream function is a good way to represent a solenoidal flow in 2D, it replaces a vector with a scalar function with no loss of information. You can define a stream function in 3D, but you need a vector function, so you gain nothing over using the velocity vector itself.

9.3   Velocity Potential

The velocity potential is simply a scalar function ϕ whose gradient is aligned to the velocity field, such that $u_i = \partial\phi/\partial x_i$, or in vector form $\vec{u} = \nabla\phi$. Thus for 2D Cartesian and cylindrical coordinate systems, the velocity potential is related to the velocity components by,

$$u = \frac{\partial\phi}{\partial x}, v = \frac{\partial\phi}{\partial y} \tag{9.3}$$

$$u_r = \frac{\partial\phi}{\partial r}, u_\theta = \frac{1}{r}\frac{\partial\phi}{\partial\theta} \tag{9.4}$$

A manifestation of the vector identity equation 6.4 on page 102 tells us that the velocity potential is a valid description of irrotational flow. Flows where both the stream function and the velocity potential are valid are termed potential flows. Along a line of $\phi(x,y) = const$, $d\phi = (\partial\phi/\partial x)dx + (\partial\phi/\partial y)dy = 0$ and from the velocity potential definition, equation 9.3, we can write $udx + vdy = 0$ along a equipotential line. Thus, the equation for a line of constant velocity potential is.

$$\frac{dy}{dx}\big|_{\phi=const} = -\frac{u}{v}. \tag{9.5}$$

9.4   Graphical Interpretation of Stream Function and Velocity Potential

Since the stream function is an expression of the volume flux, streamtubes are conditions where the volume flow is constant. When the stream function changes value hopping from one streamline to the next, the streamtube bounded by these two streamlines (in 2D) has a volume flow equal to this increment. In a similar way, contours of velocity potential may be drawn. Comparing the equations for streamlines (equation 4.1 on page 59) and equipotentials (equation 9.5 above), these lines are mutually perpendicular and a 'flow net' can be drawn.

9.5   Governing Equations for Potential Flow

The great advantage of potential flows is the simplicity of their solution. Mass conservation is assured by the solenoidal condition ($\nabla \cdot \vec{u} = 0$) and since the velocity is defined by the gradient of the velocity potential $\vec{u} = \nabla\phi$ then the solution is the Laplacian of the velocity potential, $\nabla \cdot (\nabla\phi) = \nabla^2\phi = 0$. For 2D problems the conditional of irrotationality $\partial u/\partial y - \partial v/\partial x = 0$ leads to $\partial^2\psi/\partial x^2 + \partial^2\psi/\partial y^2 = 0$ and similarly for cylindrical systems. A useful feature, discussed more fully below is that because solutions to the Laplace solutions are linear, solutions of simple flows can be added to define more complex flow fields.

9.5.1 Pressure Field and Drag and Lift Coefficients

We can obtain the pressure field in the domain directly by post-processing velocity field once it is available, from the Bernoulli equation,

$$p(x, y) = p_o - (1/2)\rho(|\nabla\phi|^2) = p_o - (1/2)\rho(u^2 + v^2) \tag{9.6}$$

and similarly cylindrical coordinates. This decoupling of the flow and the pressure field simplifies solution tremendously. This leads us to calculating the pressure coefficient at a point in the flow, or more specifically on the surface of an object, in terms of the free stream velocity and the stagnation pressure, where the stagnation pressure is expressed in terms of freestream conditions, $p_o = p_\infty + (1/2)\rho u_\infty{}^2$.

$$C_p(x, y) = \frac{p_o - p_\infty}{(1/2)\rho u_\infty{}^2} = 1 - \frac{u^2 + v^2}{u_\infty{}^2} \tag{9.7}$$

9.5.2 Boundary Conditions

If at any point in the flow we wished to find the velocity component in a given direction $\hat{n}$ then this could be obtained from the velocity potential, $\vec{u} \cdot \hat{n} = \nabla\phi \cdot \hat{n} = \partial\phi/\partial n$. A similar expression could be obtain for the stream function. In general, if we need the velocity in a certain direction, we obtain the gradient of the potential in that direction. For boundaries on the edge of the domain we can assume free stream conditions, $u = u_0, v = 0$, thus $\partial\phi/\partial x = u_0$ and $\partial\phi/\partial y = 0$. At wall boundaries, for instance at the surface of the object the velocity must be tangential to the surface, in other words the normal component is zero $\partial\phi/\partial n = 0$. For the stream function wall boundary, $\partial\psi/\partial s = 0$ where s is the arc length along the surface, although this is not often used as the boundary is defined by a streamline of given stream function value. Recall, as with all potential flows, the fluid is inviscid and thus the boundary condition at the wall is *not* the no-slip condition as defined in section 2.6.4.3 on page 26.

9.5.3 Potential Flow Solution Procedure

The solution procedure is very simple. (1) Apply appropriate boundary conditions and solve for the stream function or velocity potential, either analytically, or numerically using the Laplace equation. (2) Obtain the velocity components from the stream function or velocity potential definition and (3) Obtain the pressure field from the Bernoulli equation.

	Velocity Potential	Stream Function
Uniform	$\phi(x, y) = u_o x$	$\psi(x, y) = u_o y$
(at angle α)	$\phi(x, y) = u_o(x \cos\alpha + y \sin\alpha)$	$\psi(x, y) = u_o(y \cos\alpha - x \sin\alpha)$
	$\phi(r, \theta) = u_o r \cos\theta$	$\psi(r, \theta) = u_o r \sin\theta$
Source/Sink	$\phi(r, \theta) = \frac{\dot{V}}{2\pi} \ln r$	$\psi(r, \theta) = \frac{\dot{V}}{2\pi}\theta$
Free Vortex	$\phi(r, \theta) = -\frac{\Gamma}{2\pi}\theta$	$\psi(r, \theta) = \frac{\Gamma}{2\pi} \ln r$

Table 9.1: *Velocity Potential and Stream Functions for Baseline Potential Flows*

9.6   Baseline Potential Flow Solutions

Here three baseline cases are defined, from which more complex flows can be constructed and table 9.1 lists the stream functions and velocity potentials. Uniform flow and the free vortex have been shown previously in Fig. 7.5 on page 109.

9.6.1   Uniform Flow

The flow is defined by Cartesian coordinates and $u(x, y) = u_o$ and $v(x, y) = 0$, and from section 7.8 on page 109 the flow is both solenoidal and irrotational and has no circulation. This is a basic building block for defining the flow past objects in potential flow.

9.6.2   Point Source/Sink

Here imagine the flow emanating or disappearing radially through a very small hole on a flat surface. Using cylindrical coordinates the volume flow per unit length (out of the page) is $\dot{V} = 2\pi r u_r$, thus $u_r(r, \theta) = \dot{V}/(2\pi r)$ for a source (positive $\dot{V}$). The flow is both solenoidal and irrotational, thus is a valid potential flow.

9.6.3   Free Vortex

The characteristics of this flow have already been discussed in section 7.8 on page 109, where the velocity is defined as $u_\theta = C/r$. The constant is written in terms of the circulation (or vortex strength), $\Gamma = -2\pi r u_\theta$, thus $C = -\Gamma/(2\pi)$.

9.7   Superposition of Baseline Potential Flow Solutions

In this section we combine the various base flows defined in section 9.6 using superimposed stream functions and velocity potentials defined in table 9.1 in order to simulate flow physics that approach reality. This is an important precursor to the Joukowsky transformation in section 9.9.2 and the Kutta condition in section 9.9.3 required to predict the lift generated by 2D aerofoils. The velocity potential and stream functions are defined in table 9.2 and shown in Figure 9.1.

9.7.1   Doublet

Doublets are constructed with a sink/source pair each of volume flow $\dot{V}$ and a separation distance S, and the source sink pair are brought together such that the product $S\dot{V}$ remains constant at a value Π. In Figure 9.1 the source has been placed to the left of the sink and the direction of flow is out of the origin to the left and back in from the right. The directionality is defined by the signs on the stream function and velocity potential listed in table 9.1. In essence a doublet is a singularity that introduces a double lobed flow pattern.

9.7.2   Uniform Flow and a Source

If we define a source of strength $\dot{V}$ at the centre of a r, θ coordinate system and superimpose a uniform flow then the resulting streamlines are sketched in Figure 9.1b. We can recover the velocity components from equation 9.2 on page 118 to obtain $u_r(r, \theta) = u_o \cos \theta (\dot{V}/2\pi r)$ and $u_\theta(r, \theta) = -u_o \sin \theta$. Equating these to zero gives us the location of the stagnation point at $(r, \theta) = (\dot{V}/2\pi u_o, \pi)$, which also defines the dividing streamline which has a stream function value of $\psi(r, \theta) = \dot{V}/2$. This defines the boundary between the two flows - they do not mix because the fluid is inviscid, and is known as the *dividing*

	Velocity Potential	Stream Function
Doublet	$\phi(r,\theta) = \frac{\Pi}{2\pi}\frac{\cos\theta}{r}$	$\psi(r,\theta) = -\frac{\Pi}{2\pi}\frac{\sin\theta}{r}$
Uniform Flow and a Source	$\phi(r,\theta) = u_o r\cos\theta + \frac{\dot{V}}{2\pi}\ln r$	$\psi(r,\theta) = u_o r\sin\theta + \frac{\dot{V}}{2\pi}\theta$
Rankine Oval	$\phi(r,\theta) = u_o r\cos\theta + \frac{\dot{V}}{2\pi}(\ln\frac{r_1}{r_2})$	$\psi(r,\theta) = $ $u_o r\sin\theta + \frac{\dot{V}}{2\pi}(\theta_1 - \theta_2)$
Uniform Flow over a Circular Cylinder	$\phi(r,\theta) = u_o r\cos\theta\left(1 + \frac{\Pi}{2\pi u_o r^2}\right)$	$\psi(r,\theta) = $ $u_o r\sin\theta\left(1 - \frac{\Pi}{2\pi u_o r^2}\right)$
Uniform Flow over a Circular Cylinder with Circulation	$\phi(r,\theta) = $ $u_o r\cos\theta\left(1 + \frac{\Pi}{2\pi u_o r^2}\right) - \frac{\Gamma}{2\pi}\theta$	$\psi(r,\theta) = $ $u_o r\sin\theta\left(1 - \frac{\Pi}{2\pi u_o r^2}\right) + \frac{\Gamma}{2\pi}\ln r$

Table 9.2: *Velocity Potentials and Stream Functions for Superposed Potential Flows*

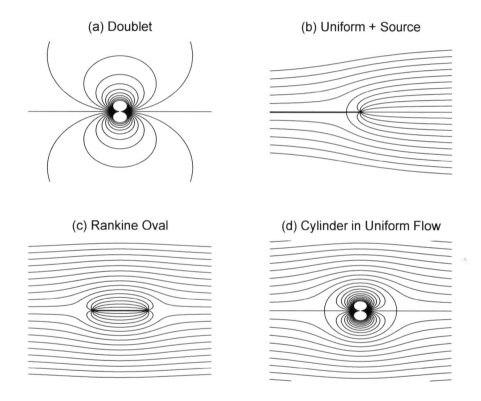

Figure 9.1: *Stream Functions for Doublet, Uniform+Source, Rankine Oval and Circular Cylinder in Uniform Flow*

streamline. Therefore, if we wanted to define the inviscid flow around a shape defined by the stream function $\psi(r,\theta) = \dot{V}/2$ we just need to define u_o and $\dot{V}$ appropriately.

9.7.3 Uniform Flow and a Source and a Sink (Rankine Oval)

We now do two things to the above configuration : (1) offset the source from the origin by a distance R towards and aligned with the direction of flow, and add a sink of the same strength at a distance R *downstream of the origin*, in the direction of flow. This create a dividing streamline that is closed (has no end) and defines a shape known as a Rankine Oval. If we define a point at (r,θ), then the distances and angles from the source and the sink locations to our point can be defined as (r_1,θ_1) and (r_2,θ_2).

As outlined in the previous section, the velocity components may be derived from either equations 9.4 or 9.2. The stagnation (zero velocity) locations can be found to be $\theta = 0$ and $\pm(R^2 + \dot{V}R/(\pi u_o))^{1/2}$. Inserting either of these locations into the general equation for the stream function above shows that the dividing streamline, defining the shape of the oval has a value zero. Again, the fluid from the uniform flow does not mix with the flow emitted by the source, which is completely consumed by the sink.

9.7.4   Uniform Flow over a Circular Cylinder

To define a cylinder of radius R we need to define this surface as a streamline. We can do this by adding a uniform flow, in r, θ coordinates to a doublet. If we set $R^2 = \Pi/(2\pi u_o)$, for $r = R$ the stream function vanishes, in other words the surface is defined by a stream function of zero. Therefore for a uniform flow over a circular cylinder, $\phi(r, \theta) = u_o r \cos \theta (1 + (R/r)^2)$ and $\psi(r, \theta) = u_o r \sin \theta (1 - (R/r)^2)$. If we use the stream function definition (equation 9.2 on page 118) to recover the velocity components we find that $u_r(r, \theta) = u_o \cos \theta (1 - (R/r)^2)$ and $u_\theta(r, \theta) = -u_o \sin \theta (1 - (R/r)^2)$, confirming that at $r = R$, $u_r = 0$.

The result is shown in Figure 9.1 where two stagnation points ($u_r = u_\theta = 0$) are located at $\theta = 0, \pi$. It is observed that the flow is symmetrical about both the vertical and the horizontal axes of the cylinder, and from equation 9.6 on page 119, so must the pressure distribution, which may be obtained from equation 9.7 on page 119. Resolving forces in the vertical direction, this means that the object generates no lift, which seems reasonable. However, the object also generates no drag, which is in direct contravention of experimental evidence, for instance figure 5.14 on page 87. This fundamental error of potential flow mystified scientists in the 18-19th centuries and was known as d'Alembert's Paradox.

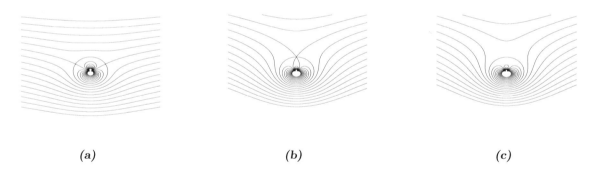

(a)   (b)   (c)

Figure 9.2: *Lifting Cylinders with (a)* $\Gamma < 4\pi R u_o$ *(b)* $\Gamma = 4\pi R u_o$ *and (c)* $\Gamma > 4\pi R u_o$

9.7.5   Uniform Flow over a Circular Cylinder with Circulation

To the uniform plus doublet flow of section 9.7.4 we now add a vortex, which introduces flow asymmetry about the horizontal plane, and thus generates lift. As shown in Figure 9.2 the vortex strength defines the flow pattern. The trivial result, $\Gamma = 0$ is shown in section 9.7.4, others include

- $\Gamma < 4\pi R u_o$: two stagnation points
- $\Gamma = 4\pi R u_o$: one stagnation point, at $(r, \theta) = (R, \pi/2)$
- $\Gamma > 4\pi R u_o$: two stagnation points, one inside and one outside the cylinder.

The radial and tangential velocities can be obtained by differentiating the stream function (or the velocity potential) to obtain,

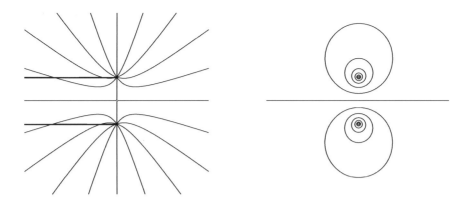

Figure 9.3: *Reflecting a Potential Flow about a Plane to simulate the Presence of a Boundary*

$$u_r = u_o \cos\theta \left(1 - \frac{R^2}{r^2}\right)$$

$$u_\theta = -u_o \sin\theta \left(1 + \frac{R^2}{r^2}\right) - \frac{\Gamma}{2\pi r}.$$

Surface velocities, where $r = R$, $u_r = 0$ (due to our boundary condition assumption, section 9.5.2) and $u_\theta = -2u_0 \sin\theta - \Gamma/(2\pi R)$. These lead to the corresponding surface pressure distribution,

$$C_p(R, \theta) = 1 - \frac{|u|^2}{u_o{}^2} = 1 - 4sin^2\theta - \left(\frac{\Gamma}{2\pi u_o R}\right)^2 - \left(\frac{2\Gamma}{\pi u_o R}\right)\sin\theta$$

Taking a elemental area of the cylinder $dA = Rd\theta$ and defining components of the unit normal in the x and y directions as $n_x = \cos\theta$ and $n_y = \sin\theta$ we can calculate the total dimensional drag and lift force for the cylinder,

$$F_D = 0, F_L = \rho u_o \Gamma \tag{9.8}$$

The lift force defined is known as the Kutta-Joukowsky Theorem and shows that the lift is defined by the amount of circulation of the vortex flow. It is also known as the Magnus effect and for spherical geometries is why golf, tennis, cricket and foot balls curve in flight if they spin. This is valid for *any* 2D body of any shape, hence the value of potential flow for aerodynamics problems - the problem simplifies to defining a representative shape (the Joukowsky transformation) and the correct amount of circulation to give a realistic flow field (the Kutta condition).

9.8   Imposing Plane Boundaries

Until now we have examined unbounded potential flows, where the edges of the domain have no gradients. Often we wish to examine the flow around objects, or examine the flow change due to the presence of a flat planar surface. This can be achieved by reflecting (imaging) the flow feature about the desired boundary plane. Figure 9.3 shows a source and a free vortex reflected about the $y = 0$ plane such that no flow crosses that plane. The principle can also be applied to superposed flows to simulate the effect of near ground interactions on lift. The principle can also be applied to a pair of parallel walls, but then a series of reflections is required.

9.9   Analytical 2D Aerofoil Theory

Here it is shown how to use a coordinate transform to map the flow around a lifting cylinder (as discussed in section 9.7.5) to the flow around an aerofoil like shape. Although we know the lift that can be generated by virtue of the Kutta-Joukowsky Theorem (equation 9.8) directly without doing the transformation, we need the aerofoil shape to define the Kutta condition for that shape. Therefore, first we define the mathematical environment, then the transformation, then the Kutta condition.

9.9.1   Complex Representation of Potential Flow

In mathematics, a conformal map is a function which preserves angles locally. This reflects very nicely the orthogonality of the stream and the velocity potential functions in potential flow. Thus, if we can transform a potential flow solution from one geometry (domain) to another, using a conformal mapping method, then the local orthogonality between the streamlines and the equipotentials will be preserved.

A complex number z is a sum of a real and imaginary part such that $z = x + iy$ where $i = \sqrt{-1}$ and $1/i = i/i^2 = -i$. Therefore the complex number defines a point on a 2D x-y plot, and is in essence a way of packing two numbers into one. It can also be written in polar form such that $z = re^{i\theta} = r(\cos\theta + i\sin\theta)$ where $r^2 = x^2 + y^2$. Likewise a complex function $f(z) = f(x+iy) = f_1(x,y) + if_2(x,y)$ can be composed of two real functions, and in the case of potential flow, the real part of $f(z)$ is the velocity potential and the imaginary part (ie the real coefficient of the imaginary part) is the stream function, such that $f(z) = \phi(x,y) + i\psi(x,y)$. There are a number of reasons why this extra complexity (no pun intended) is worthwhile :

- Both the real and the imaginary parts of any arbitrary function of the complex variable z are solutions to the Laplace equation.
- Contours of the real and imaginary parts of the complex function (which is stressed, are both real!) are always orthogonal to each other.
- To find the velocity components, we can differentiate $\phi + i\psi$ directly, with respect to x or iy, or the complex function directly, with respect to z, as shown below in equation 9.9.

$$\frac{d}{dz}f(z) = u - iv \qquad \frac{d}{dx}(\phi + i\psi) = u - iv \qquad \frac{d}{dy}(-i\phi + \psi) = iu + v \qquad (9.9)$$

The baseline potential flows listed in section 9.6 on page 120 and defined in table 9.1 can all be represented using complex functions $f(z)$. For example for uniform flow, $u(x,y) = u_o$ and $v(x,y) = 0$, giving $\phi(x,y) = u_o x$, $\varphi = u_0 y$, thus $f(z) = u_o z$ and $df/dz = u_o = u - iv$. In polar form, and for an incline α, $f(z) = rze^{-i\alpha}$. Similarly, complex potentials for the other baseline cases are a source/sink z_o from the origin, $f(z) = (\dot{V}/(2\pi))\ln(z - z_o)$ and free vortex $f(z) = -(i\Gamma/(2\pi))\ln(z - z_o)$ doublet, $w(z) = \Pi/(2\pi(z - z_o))$.

As an example, consider the complex function $f(z) = Az^n$, where A, n are constants. We can see that $n = -1$ would give us a doublet field. If $n = 2$, then $z^2 = (x+iy)^2 = A(x^2 + y^2) + i2Axy$. Therefore $\phi = A(x^2 + y^2)$ and $\psi = 2Axy$. Differentiating, $d/dz(f(z)) = 2Az = 2A(x + iy)$. Thus $u = 2Ax$ and $v = -2Ay$, the equation of a hyperbola, and this flow represents a 90 deg corner flow, as shown in Figure 9.4. Likewise $n = 3, 1.5, 2/3, 1/2$ represent flow around other angles. These baseline complex potentials may be superimposed as previously, and the complex potential for a cylinder with rotation for example is given by,

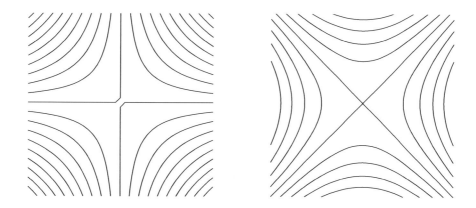

Figure 9.4: *Complex Potential Corner Flow showing Streamlines (L) and Equipotentials (R)*

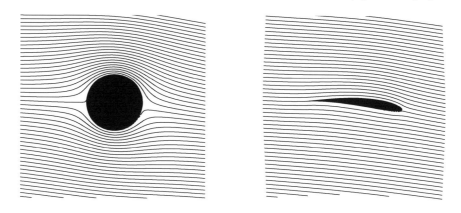

Figure 9.5: *The Jowkowsky Transformation for Aerofoil Construction from the z_1 (L) to the z_2 plane.*

$$f(z) = u_o \left(z + \frac{R^2}{z} \right) + i \frac{\Gamma}{2\pi} \ln(z) \qquad (9.10)$$

9.9.2 Conformal Mapping and the Joukowsky Transformation

Much like you can convert velocities from Cartesian to cylindrical coordinate systems, we can transfer complex numbers and functions from one coordinate system to another. For instance the complex potential for uniform flow can be transformed from (x, y) to (r, θ). Generally we convert a number in one plane, ie $z_1 = x + iy$ to another $z_2 = \xi + i\zeta$ using some transformation function. In aerodynamics the most common is the Joukowsky function, $z_2 = z_1 + b^2/z_1$. The base transformation, when $b = R$ maps the cylinder in the z_1 plane to a flat line in the z_2 plane, and more realistic cases are where the Joukowsky transformation is applied to a cylinder offset along the x axis in the -ve direction and then again offset in the positive y direction. These generate symmetrical and cambered aerofoils respectively, the latter shown by Figure 9.5. Because of the properties of complex functions noted above in section 9.9.1, the tranformed flowfield is a valid flowfield for the transformed object (recall : the transformed object is just another streamline - part of the flowfield).

9.9.3 Kutta Condition

Recall from section 9.7.5 that it is the circulation that is effectively an arbitrary constant in defining lift from the point of view of flow over an aerofoil. This means we require some way to match the circulation to a realistic flow around any given aerofoil at any given angle of attack in order to calculate the lift using the Kutta-Joukowsky Theorem (equation 9.8). This is achieved by imposing the Kutta condition on the flow defined by equation 9.10, such that a body with a sharp trailing edge which is moving through a fluid will create about itself a circulation of sufficient strength to hold the rear stagnation point at the trailing edge. In other words the streamlines exiting the trailing edge of the aerofoil. This can be identified on Figure 9.5 where the stagnation point sits on the $y = 0$ line in the z_2 plane.

9.10   Summary and Further Reading

A brief introduction to potential flow has been given which built on the understanding of kinematics given in chapter 7 on page 105. Good sources for further reading are older fluids texts, the one by Currie is recommended. Panel methods are the next step and permit general 2D and 3D potential flows to be calculated numerically.

10 | Compressible Fluid Flow

10.1 Introduction

Here we analyse the unexpected properties of a compressible fluid when it nears and exceeds the Mach number, first introduced on page 116. It is relevant for most industrial gas systems and aircraft flight. In the compressible fluid conservation equations on page 115 the density is present in the conservation equations of mass, momentum and energy, and is also defined by the gas law. Temperature occurs in the energy equation and the equation of state. Likewise the pressure is found in both the momentum equation and the equation of state. This means the equations are tightly coupled with many interdependent variables. For this reason the majority of this chapter is restricted to one-dimensional compressible flow. It is an expression of the true linking of the subjects of thermodynamics and fluid mechanics.

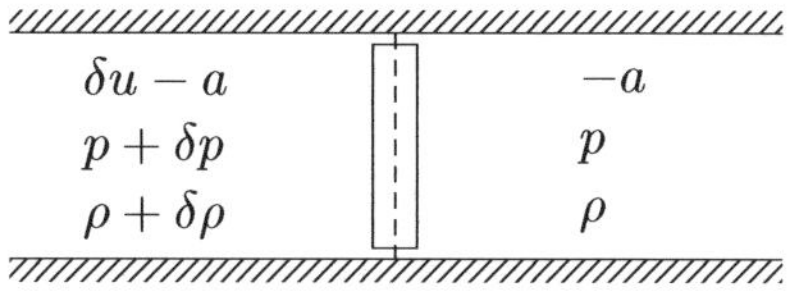

$\delta u - a$		$-a$
$p + \delta p$		p
$\rho + \delta \rho$		ρ

Figure 10.1: *Disturbed (L) and undisturbed (R) fluid either side of a pressure wave*

10.2 Compressible Bernoulli Equation

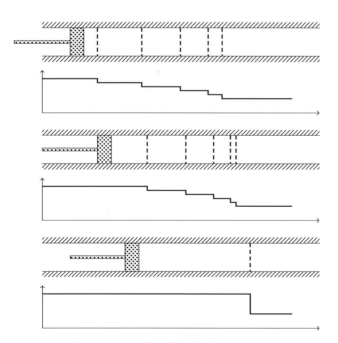

Figure 10.2: *Formation of a Shock due to the superposition of compression waves.*

We start from the steady flow energy equation (equation 4.6 on page 64) with no heat or work transfers and no potential energy changes, that is given a stagnation state 0 and two other states 1 and 2 then $e_{h,0} = e_{h,1} + u_1^2/2 = C_p T_2 + u_2^2/2$. Differentiating this along the streamline direction gives $dh + u\,du = 0$. Using one of the TdS equations (eqn 2.21, page 40) and, providing the flow is isentropic we obtain the compressible Bernoulli equation,

$$\frac{dp}{\rho} + udu = 0. \qquad (10.1)$$

Note this equation was obtained from the total energy equation, whereas the incompressible Bernoulli equation (eqn 4.3, page 61) was derived by integrating a force balance along a streamline and only considers mechanical energy. Note that for the integral form, a *total* energy balance, $\int dp/\rho + u^2/2 = const$ requires a pressure-density dependence to be defined. This by virtue of the gas law, involves temperature.

10.3 Speed of a Pressure Wave in Still Fluid

Pressure waves, of which sound waves are an example, transmit molecular kinetic energy through space by molecular collisions and a very important parameter for compressible flow is the speed of the pressure wave propagation relative to the local speed of the fluid, the Mach Number. Let us take a long insulated tube filled with a still inviscid fluid and we tap one end. An instantaneous small pressure pulse from left-hand side propagates at speed a down the pipe. As shown in Fig. 10.1 we take our reference about the wave. Since the pressure pulse is a discontinuity we cannot apply differential conservation equations but we can apply integral forms over the discontinuity. Therefore the mass conservation equation, for equal areas becomes $(\rho u)_{lhs} = (\rho u)_{rhs}$ leading to $(\rho + \delta\rho)\,\delta u = a\delta\rho$. Likewise momentum conservation becomes $\delta u = \frac{\delta p + a^2 \delta\rho}{2a(\rho+\delta\rho)}$. Equating these two gives $a_0 = \sqrt{\partial p/\partial \rho}$. Recall that we assumed still fluid, thus this represents the wave speed in still fluid hence the subscript denoting stagnation conditions. Since the pressure pulse was small, the compression process is approximately adiabatic therefore $p/\rho^\gamma = Const$ and thus $a_0 = \sqrt{\gamma R T_0}$.

Also since $p/\rho^\gamma = p_0/\rho_0^\gamma$, $a = a_0\,(\rho/\rho_0)^{(\gamma-1)/2}$, pressure waves travel faster in denser fluids. We could also derive an energy balance from the steady flow energy equation (eqn 4.5 on page 64) assuming no work or heat transfers which gives $0 = \rho u\left\{e_h + \delta e_h + 1/2\,(u+\delta u)^2 - \left(e_h + 1/2 u^2\right)\right\}$. Eliminating δu gives $\delta e_h - \delta p/\rho = 0$. With reference to the TdS equations, (eqn 2.21, page 40) we see that $T\delta s = 0$, showing that small pressure disturbances are isentropic processes.

10.4 Isentropic Flow Equations

Starting from the definition of stagnation enthalpy $e_{h,0} = e_{h,1} + u_1^2/2$ we can define first the relationship between temperature and Mach Number, $T_0 = T_1\left(1 + \frac{\gamma-1}{2}\mathrm{Ma}_1^2\right)$ and from this other equations in terms of wave speed, density and pressure,

$$a_0 = a_1\left(1 + \frac{\gamma-1}{2}\mathrm{Ma}_1^2\right)^{\frac{1}{2}}, \qquad \rho_0 = \rho_1\left(1 + \frac{\gamma-1}{2}\mathrm{Ma}_1^2\right)^{\frac{1}{\gamma-1}}, \qquad p_0 = p_1\left(1 + \frac{\gamma-1}{2}\mathrm{Ma}_1^2\right)^{\frac{\gamma}{\gamma-1}}.$$

For the special case of unit Mach Number we have the critical flow equations,

$$\frac{T_1}{T_0} = \frac{2}{\gamma+1}, \qquad \frac{p_1}{p_0} = \left(\frac{2}{\gamma+1}\right)^{\frac{\gamma}{\gamma-1}}, \qquad \frac{\rho_1}{\rho_0} = \left(\frac{2}{\gamma+1}\right)^{\frac{1}{\gamma-1}}, \qquad \frac{a_1}{a_0} = \left(\frac{2}{\gamma+1}\right)^{\frac{1}{2}}.$$

These are constants for a given fluid and are often used in calculations. For air they are,

$$\frac{T^*}{T_0} = 0.8333, \qquad \frac{p^*}{p_0} = 0.5283, \qquad \frac{\rho^*}{\rho_0} = 0.6339, \qquad \frac{a^*}{a_0} = 0.9129.$$

10.5 Compression and Expansion Waves

Consider an insulated duct in which a piston is at one end and accelerates in discrete jumps as shown in Fig. 10.2. The piston speeds up from $u = u(1)$ to $u = u(5)$, a terminal velocity in finite steps. A series of compression waves pass down the channel at different speeds because the piston is compressing the gas. The gas is most compressed near the piston hence the compression wave is fastest here. The initial compression wave travelling down the tube is travelling the slowest because in front of this wave is the undisturbed fluid. This means the faster later compression waves catch up with the slower initial waves and the small pressure intervals on each compression wave add to form a shock wave.

10.6   Normal Shock Waves

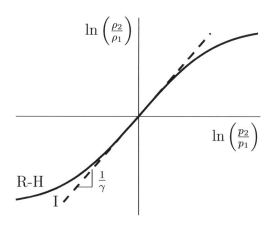

Figure 10.3: *Rankine-Hugoniot and Isentropic Pressure-Density Curves.*

As with the single pressure pulse of section 10.3, to conserve properties across a shock, we must use integral equations, therefore mass, $\rho_1 u_1 = \rho_2 u_2$, momentum, $\rho_1 u_1^2 + p_1 = \rho_2 u_2^2 + p_2$ and energy, $\frac{1}{2}u_1^2 + \frac{\gamma}{\gamma-1}\frac{p_1}{\rho_1} = \frac{1}{2}u_2^2 + \frac{\gamma}{\gamma-1}\frac{p_2}{\rho_2}$. We also assume adiabatic (but not reversible, hence not isentropic) conditions and also a perfect gas. Since we have 3 equations and 6 variables we solve for ratios and the Rankine-Hugonoit equations are

$$\frac{u_1}{u_2} = \frac{\rho_2}{\rho_1} = \frac{1 + (\gamma+1)/(\gamma-1)\,(p_2/p_1)}{(\gamma+1)/(\gamma-1) + p_2/p_1} \tag{10.2}$$

Figure 10.3 shows equation 10.2 versus the isentropic compression relation (section 2.9.8.4 on page 39). It can be seen that weak shocks are nearly isentropic. It can also be seen that there appear to be two valid quadrants for the Rankine-Hugonoit equations. In fact there is only one, since the entropy change across a shock may be written as $\frac{S_2-S_1}{C_v} = \ln\left(\frac{p_2}{p_1}\right) - \gamma\ln\left(\frac{\rho_2}{\rho_1}\right)$. Then, by using equation 10.2 we can show that a positive entropy change only occurs when $\rho_2/\rho_1 \geq 1$ since $\left(\frac{S_2-S_1}{C_v}\right)_{R-H} = \gamma\left(\ln\left(\frac{\rho_2}{\rho_1}\right)_I - \ln\left(\frac{\rho_2}{\rho_1}\right)_{R-H}\right)$. From continuity $u_1/u_2 \geq 1$, therefore fluids slow and compress across shocks. The Rankine-Hugonoit equations may also be expressed as a function of upstream Mach Number in terms of downstream Mach Number (equation 10.3), density ratio (equation 10.4) and pressure ratio (equation 10.5). These are shown graphically on Fig. 10.4.

$$\text{Ma}_2^2 = \frac{1 + [(\gamma-1)/2]\,\text{Ma}_1^2}{\gamma\text{Ma}_1^2 - (\gamma-1)/2} \tag{10.3}$$

$$\frac{\rho_2}{\rho_1} = \frac{(\gamma+1)\,\text{Ma}_1^2}{(\gamma-1)\,\text{Ma}_1^2 + 2} \tag{10.4}$$

$$\frac{p_2}{p_1} = 1 + \frac{2\gamma}{(\gamma+1)}\left(\text{Ma}_1^2 - 1\right) \tag{10.5}$$

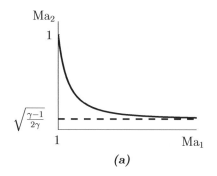

(a)

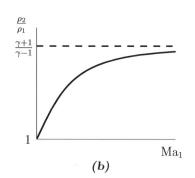

(b)

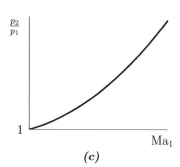

(c)

Figure 10.4: *Normal Shock Characteristics.*

10.7   Duct Flow: The Influence Coefficients

Here we examine how changes in duct area, and the effect of heat addition/rejection and friction affect the properties of a compressible fluid. Fluid variables include the speed u, density ρ, pressure

p, temperature T, Mach Number Ma, frictional force f, heat transfer q. Generally an duct element of length δx and area δA subjected to a force δf and heat addition δq. We assume steady flow and that there are no shocks. This allows us to use the derivative forms of the conservation equations. Expanding the mass conservation equation, $\frac{d(\rho u A)}{dx} = 0$ gives $uA\frac{d\rho}{dx} + \rho u\frac{dA}{dx} + \rho A\frac{du}{dx} = 0$ and dividing by $\frac{\rho u A}{dx}$ gives

$$\frac{d\rho}{\rho} + \frac{du}{u} + \frac{dA}{A} = 0. \tag{10.6}$$

The differential momentum balance is $\rho u\frac{\delta u}{\delta x} = -\frac{\delta p}{\delta x} + \frac{\delta f}{\delta x}$, dividing by $\frac{p}{dx}$ gives $\frac{\rho}{p}u\,du = -\frac{dp}{p} + \frac{df}{p}$. Since $a^2 = \frac{dp}{d\rho}$ then

$$\gamma Ma^2\frac{du}{u} + \frac{dp}{p} = \frac{df}{p}. \tag{10.7}$$

The local enthalpy, kinetic energy and heat transfer are related by $e_h + \frac{u^2}{2} = q$. In differential form $C_p\frac{\delta T}{\delta x} + u\frac{\delta u}{\delta x} = \frac{\delta q}{\delta x}$, dividing by $\frac{C_p T}{dx}$ gives $\frac{dT}{T} + \frac{u\,du}{C_p T} = \frac{dq}{C_p T}$ and finally

$$\frac{dT}{T} + (\gamma - 1)\,Ma^2\frac{du}{u} = \frac{dq}{C_p T} \tag{10.8}$$

For a perfect gas $p = \rho RT$ and in differential form $\frac{1}{\rho T}\frac{\delta p}{\delta x} - \frac{p}{\rho^2 T}\frac{\delta \rho}{\delta x} - \frac{p}{\rho T^2}\frac{\delta T}{\delta x} = 0$. Dividing by $\frac{p}{\rho T\,dx}$ gives,

$$\frac{dp}{p} - \frac{d\rho}{\rho} - \frac{dT}{T} = 0. \tag{10.9}$$

Equations 10.6, 10.7, 10.8 and 10.9 are 4 relations of du, dp, dT and $d\rho$ in terms of u, ρ, p, T, Ma, f, q and A.

The factors $\frac{dA}{A}$, $\frac{df}{p}$ and $\frac{dq}{C_p T}$ are known as the 'influence coefficients' and define how much the flow is affected by external influences. The power of the analysis comes from eliminating certain factors from equations 10.6–10.9.

10.7.1 Varying Area – Frictionless Fluid – No Heat Transfer

This is achieved by eliminating $\frac{d\rho}{\rho}$ and setting $\frac{df}{p} = \frac{dq}{C_p T} = 0$. The result is $\frac{du}{u} = \frac{1}{Ma^2 - 1}\frac{dA}{A}$. This is a key result. So when Ma < 1 when $dA/A > 0$ then $du/u < 0$, i.e. the velocity decreases. This is what we would expect from incompressible mass conservation. However notice that when Ma > 1 now when $dA/A > 0$ then $du/u > 0$. I.e. as the area increases so does the velocity.

10.7.2 Constant area duct with heat addition, no friction

This is achieved by setting $\frac{df}{p} = \frac{dA}{A} = 0$. The result is $\frac{dT}{T} = \frac{\gamma Ma^2 - 1}{Ma^2 - 1}\frac{dq}{C_p T}$. Since $\gamma > 1$ always, therefore if $dq > 0$ this increases T for Ma $< \frac{1}{\gamma^{1/2}}$ and Ma > 1. Likewise for $dq > 0$ this decrease T for $1 > $ Ma $> \frac{1}{\gamma^{1/2}}$. Plotted graphically, the relations are known as Rayleigh Lines.

10.7.3 Constant area duct with friction, no heat addition

The result is $\frac{dT}{T} = \frac{-(\gamma-1)Ma^2}{Ma^2 - 1}\frac{df}{p}$. Since $\gamma > 1$ when $df > 0$ this increases T when Ma < 1 as we would expect. Conversely, when $df > 0$ decreases T when Ma > 1. Plotted graphically, the relations are known as Fanno Lines.

10.8 Duct Flow: Choking

For any duct of an inviscid fluid the mass flow down that duct is defined by, $\dot{m} = \rho(x)u(x)A(x) = \rho A\left(2C_p\left(T_0 - T\right)\right)^{1/2}$. This relates the mass flow rate to the difference between stagnation ($u = 0$) and local ($u/= 0$) conditions, here using temperature, but also pressure and density may also be used via the adiabatic relations $\frac{T}{T_0} = \left(\frac{p}{p_0}\right)^{\frac{\gamma-1}{\gamma}}$, $\frac{\rho}{\rho_0} = \left(\frac{p}{p_0}\right)^{\frac{1}{\gamma}}$. If we group the stagnation properties on the LHS we obtain the non-dimensional mass flow in terms of the local pressure ratio and duct area.

$$\frac{\dot{m}}{\rho_0 \left(2C_pT_0\right)^{1/2}} = A\left(\frac{p}{p_0}\right)^{\frac{1}{\gamma}}\left(1 - \left(\frac{p}{p_0}\right)^{\frac{\gamma-1}{\gamma}}\right)^{1/2}.$$

By plotting A against p/p_0 we get the area profile along the nozzle axis required for a converging-diverging nozzle, as shown in Figure 10.5. Note that the air enters the nozzle at near atmospheric pressure ($p/p_0 \sim 1$) and is accelerated subsonically to the throat, where the fluid can become sonic. For sonic conditions, where the critical pressure is defined we can also define a critical area, A^*, which is the throat area. As noted in section 10.7.1, if the fluid is supersonic in a diverging nozzle it will continue to accelerate and the pressure will continue to reduce. Note the Ma number at the throat cannot be more than unity, thus the maximum mass flow is defined by sonic conditions at the throat. Thus the nozzle maximum mass flow is defined only by fluid properties, throat area and stagnation conditions.

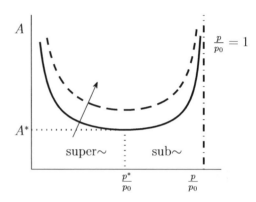

Figure 10.5: Pressure Ratio versus Nozzle Area

10.9 Converging Nozzle Flow

Consider a typical converging nozzle discharging air to the atmosphere. Pressures at three locations define the operation. The upstream stagnation pressure used to drive the air through the nozzle (p_0), the pressure at the throat (p_t) and the pressure of the air into which the nozzle flow is being delivered, often called the back pressure (p_b). The pressure difference $p_0 - p_b$ drives the flow. As shown in Figure 10.6, three nozzle states are possible. As p_b is reduced from p_0 the flow will increase in speed. It will remain subsonic throughout until $p_t = p_t^*$, when the throat goes sonic and the nozzle is choked. At this point the flow upstream of the throat in decoupled from the flow downstream of the throat, because downstream pressure waves (for instance caused by changes to p_b) cannot pass the throat. This means that as p_b is further reduced the upstream flow is unchanged. The downstream flow has to expand supersonically to get down from p_t to p_b and this is discussed in section 10.11 on page 132.

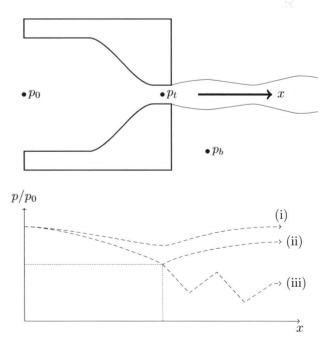

Figure 10.6: Pressure Ratio in Converging Nozzle flow with variable back pressure.

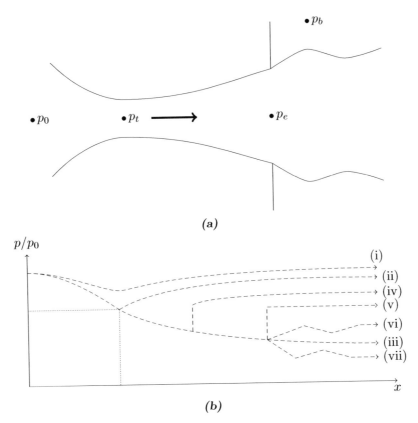

Figure 10.7: *Pressure Ratio in Converging-Diverging Nozzle flow with variable back pressure.*

10.10   Converging-Diverging (de Laval) Nozzle Flow

To deliver a supersonic flow we require a converging section to accelerate the subsonic fluid to sonic conditions and then a diverging section to accelerate it supersonically. Compared to the converging nozzle we require one additional pressure, at the exit of the diverging section p_e, as shown in Figure 10.7. A range of conditions are possible.

1. $\dot{m}^*/\dot{m} < 1$, subsonic, isentropic relations.
2. subsonic diffuser, $A_e/A_t = A_e/A^*$, sonic throat, $= 1$.
3. supersonic diffuser, p_b/p_0 correct for A_e/A^*, design condition for rockets, supersonic windtunnel.
4. $p_b|_{ii} > p_b > p_b|_{iii}$, throat remains choked, second shock forms, subsonic exit flow, pressure rise.
5. as 4
6. $p_b > p_b|_{iii}$ (over expanded), second shock cannot occur in the nozzle, supersonic exit, complex shocks form to get $p_e \rightarrow p_b$.
7. $p_b < p_b|_{iii}$ (under expanded), as above but pressure difference is the reverse of 6, complex shocks form.

10.11   Two Dimensional supersonic flow

Here we consider the flow characteristics for small and significant deflections in flow direction, due to the presence of walls, assuming the fluid is inviscid. This is an approximation since we know that viscous forces are always important near walls.

10.11.1   Isentropic compression and expansion turns

As shown in Figure 10.8 the supersonic flow is always parallel to the wall, hence the flow turn is abrupt on the line emanating from the wall turning through an angle $d\theta$. If the angle of the turn is small then, approximately $\frac{du}{dv} = \tan\beta$. If $d|u| \sim du$ and $d\theta \sim -dv/|u|$ then, $\frac{d|u|}{|u|} = -\tan\beta\, d\theta$. Since $90 \geq \beta \geq 0$, $tan\beta$ always $+ve$, so when $d\theta < 0$ (wall turns away from flow) the flow accelerates and when $d\theta > 0$ (wall turns into the flow) the flow decelerates. Using the compressible Bernoulli equation (equation 10.1 on page 127) $\frac{d|u|}{|u|} = -\frac{dp}{\rho|U|^2} = -\frac{dp}{\gamma \mathrm{Ma}^2 p}$ and knowing that $\tan\beta = \frac{1}{\left(\mathrm{Ma}^2 - 1\right)^{1/2}}$ then $\frac{\delta p}{p} = \frac{\gamma \mathrm{Ma}^2 \delta\theta}{\left(\mathrm{Ma}^2 - 1\right)^{1/2}}$. We see that for $d\theta < 0$ this causes the pressure to reduce. Therefore, turning the wall into the flow is known as a 'compression turn' while turning the wall away from the flow is known as an 'expansion turn'. The dependence of Mach number change on deflection angle can be derived but the integral form of the above expression is known is the Prantl-Meyer Function and is shown in Figure 10.9.

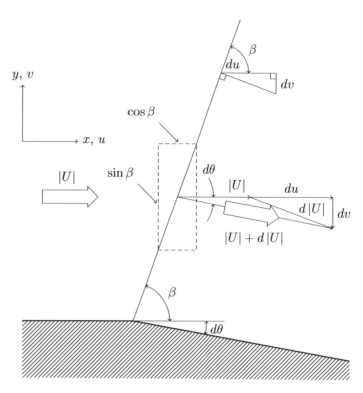

Figure 10.8: Nomenclature for 2D supersonic flow

$$\nu(\mathrm{Ma}) = \left(\frac{\gamma+1}{\gamma-1}\right)^{1/2} \tan^{-1}\left\{\left(\frac{(\gamma-1)\left(\mathrm{Ma}^2-1\right)}{\gamma+1}\right)^{1/2}\right\} - \tan^{-1}\left\{\left(\mathrm{Ma}^2-1\right)^{1/2}\right\}.$$

The difference in the two functions is used to define the Mach number change for a given turn $\theta_T = \int_{\mathrm{Ma}_1}^{\mathrm{Ma}_2} \nu(\mathrm{Ma})\, d\mathrm{Ma} = \nu(\mathrm{Ma}_1) - \nu(\mathrm{Ma}_2)$. Note this relation holds for expansion turns of any angle but only for compression turns of small angle, known as weak shocks. For significant compression turns a (strong) shock forms and the following is required.

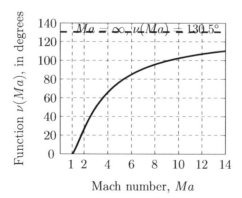

Figure 10.9: Prantl-Meyer Function for Supersonic expansion turns

10.11.2 Oblique Shocks

An oblique shock wave is a shock wave *not* normal to the flow, We can decompose into normal and tangential components as shown in Figure 10.10. From geometry we can show the upstream normal component is $u_{1n} = u_1 \sin\beta$ and the downstream normal component, $u_{2n} = u_2 \sin(\beta - \theta)$. We can

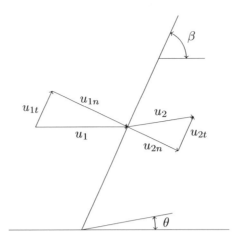

Figure 10.10: *Nomenclature for Oblique Shocks*

also define normal Mach numbers, $\mathrm{Ma}_{n1} = \frac{u_{n1}}{a_1} = \mathrm{Ma}_1 \sin\beta$, $\mathrm{Ma}_{n2} = \frac{u_{n2}}{a_2} = \mathrm{Ma}_2 \sin(\beta-\theta)$. In oblique shocks $u_{t1} = u_{t2}$ since the pressure along the shock is constant. For a shock to exist, Ma_1 and Ma_{1n} *must* be supersonic, Ma_{2n} must be subsonic and Ma_2 *may* be subsonic. Note $\mathrm{Ma}_{1t} = \mathrm{Ma}_{2t}$ since the wave speeds either side of the shock are not the same. We apply the same control volume analysis as for normal shocks (section 10.6 on page 129), but with tangential momentum. The continuity equation is $\rho_1 u_{1n} = \rho_2 u_{2n}$, normal and tangential momentum $p_1 + \rho_1 u_{1n}^2 = p_2 + \rho_2 u_{2n}^2$ and $\rho_1 u_{1n} u_{1t} = \rho_2 u_{2n} u_{2t}$. Finally the energy equation is defined $e_{h,1} + \frac{u_{1t}^2}{2} + \frac{u_{1n}^2}{2} = e_{h,2} + \frac{u_{2t}^2}{2} + \frac{u_{2n}^2}{2}$. These can be solved to obtain the shock wave angle β in terms of the upstream Mach number and the flow deflection angle,

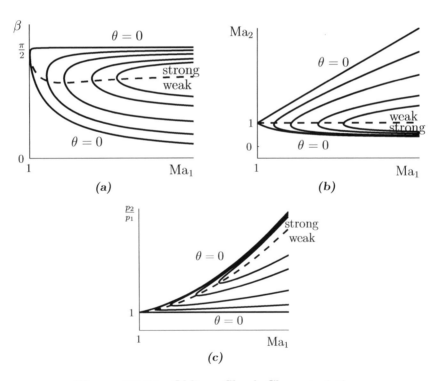

Figure 10.11: *Oblique Shock Characteristics.*

$$\mathrm{Ma}_1^2 = \frac{2\cos(\beta-\theta)}{\sin\beta\left[\sin(2\beta-\theta) - \gamma\sin\theta\right]}.$$

Two valid values of β are valid, and these correspond to the deflection angles for strong and weak shocks. An oblique shock is said to be strong if $\mathrm{Ma}_2 < 1$. Further relations for the Mach number and pressure ratios are given by

$$\text{Ma}_2^2 = \frac{1 + \left[(\gamma - 1)/2\right] \text{Ma}_1^2 \sin^2 \beta}{\sin^2 (\beta - \theta) \left[\gamma \text{Ma}_1^2 \sin^2 \beta - (\gamma - 1)/2\right]}$$

and

$$\frac{p_2}{p_1} = 1 + \frac{2\gamma}{\gamma + 1} \left(\text{Ma}_1^2 \sin^2 \beta - 1\right).$$

And all three relations are shown graphically in figure 10.11.

10.12 Summary and Further Reading

A concise introduction to 1D and 2D compressible subsonic and supersonic compressible flow is presented. Further details are available in the book by Currie.

11 | Practical Thermodynamics

11.1 Introduction

In the introductory chapters of this book (chapters 2 to 5) the emphasis was on understanding the basic physical concepts, such as individual compression/expansion and heat transfer processes. Work, heat transfer and internal energy change was considered in section 2.9.8 on page 38, and entropy change in section 2.9.11 on page 41. We then considered how sequences of these processes, incorporated into model engine cycles exchange work, heat and entropy with the environment with a given efficiency in chapter 3 on page 44. This gave us a complete picture, with the key simplification being the assumption that our working fluid was a pure perfect gas.

In this chapter we relax this assumption, and modify our knowledge to account for these realities. We also analyse several common types of process equipment that take advantage of these non-perfect fluid properties such as steam turbines and air conditioning units.

11.2 Single Phase Non-Perfect Gasses

In section 2.7 on page 29 we made use of a simple kinetic theory model of a gas, where molecules were not assumed to interact with each other and possessed negligible volume. This complete but simplified model gave us fundamental description of molecular interactions gave us directly the ideal gas law (equation 2.17 on page 32) and the gas constant, the first law of thermodynamics, constant gas specific heats and also a molecular level view of what entropy is. This is actually a very good approximation to most gases under reasonable conditions, for instance air at normal temperature and pressure.

11.2.1 Ideal Gas Mixtures : Partial Pressure and Dalton's Law

Say we have a box of volume V containing a gas mixture at a pressure p and temperature T. The total number of moles of gas in the box is n, and each of the i components have n_i such that $n = \sum n_i$. The *partial* pressure is what the total pressure would be if only the ith component was in the box. Therefore, from equation 2.17 on page 32, the partial pressure of the ith component p_i is,

$$p_i = \frac{n_i R_u T}{V} \tag{11.1}$$

and Dalton's Law states that the total pressure in the box is the sum of these partial pressures.

$$p = \sum p_i \tag{11.2}$$

Notice the equivalence between the mole fractions and the partial pressures, $n_i/n = p_i/p$ since $p = \sum p_i$ and $n = \sum n_i$.

11.2.2 Semi-Perfect Gases

Recall our definition of a perfect gas : (1) Obeys the Ideal Gas Law, equation 2.17 on page 32 and (2) Constant Specific Heats. Semi-perfect gases relax the second assumption, but require that the specific heats are a function of temperature only. The ideal gas law, and all standard thermodynamic relations still apply.

11.2.3   Real Gases

Kinetic theory (section 2.7.1 on page 30) made three key assumptions (1) collisions were perfectly elastic (2) the volume of the molecules relative to the container volume is negligible and (3) the intermolecular interactions are negligible. Of these (1) has to be true, otherwise pressure would reduce over time, however (2) and (3) are only good approximations for 'low' pressure and 'high' temperature. In fact, without kinetic theory the relationship $p/\rho = f(T)$ must be obtained empirically and several relations exist, one of the more popular is the Van De Waals' equation.

$$\left(p + a/v^2\right)\left(v - b\right) = RT \tag{11.3}$$

The empirical coefficient a accounts for attractive forces and b takes care of the volume of the molecules themselves. For everyday engineering use, the Van de Waal's equation is not often used because all subsequent thermodynamic relationships must be modified (as shown in section 11.2.4), instead the ideal gas law is modified using a compressibility factor, such that $pv = ZRT$. The compressibility factor is typically a function of pressure and temperature and are available as charts and incorporated into computer programs.

11.2.4   Throttling, and the Joule-Thompson Effect

Throttles are small steady flow devices the restrict flow and cause a significant drop in pressure, and can be a valve or orifice. The expansion valve in a domestic refrigerator, part of the reversed heat pump introduced in section 3.7 on page 47 is a classic everyday example. If a perfect gas was assumed, when a gas is pumped slowly through a flow restriction in an insulated pipe, the steady flow energy equation (equation 4.5 on page 64) the enthalpy of the gas should not change, and since $e_h = C_p T$, then neither should the temperature. Careful measurements of most gases reveal that it does, and therefore is a result of non-perfect gas behaviour. This can be revealed by the equation for the internal energy derived from equation 11.3 and the fundamental definition for internal energy, $de_u = (\partial e_u/\partial T)|_v dT + (\partial e_u/\partial v)|_T dv$. This gives $de_u = C_v dT + (a/v^2)dv$, and a similar expression can be obtained for C_p. *Generally*, this means throttling produces cooling, and the Joule-Thompson effect is widely used for refrigeration and liquefaction processes.

11.3   Pure Substances with Multiple Phases

Thus far we have only considered a fluid with one phase, here we extend thermofluid analysis to processes with multiple phases. This is important because the most important stationary power generation cycle, the Rankine (steam) cycle relies on the phase change of water. Immediately, we extend the principle of quasi-equilibrium processes and the two property rule (sections 2.9.5 and 2.9.1 on page 37 and 36 respectively) to phase change. In other words, in a sealed box, containing some liquid water and some vapour at steady state, a dynamic equilibrium exists and the rate of evaporation equals the rate of condensation. Change the pressure and/or temperature the water vapour partial pressure changes and a new stable equilibrium is set up and so on. We now analyse phase equilibria for water. Phase change is associated with an energy change. The latent heat of fusion is the energy associated with solid-liquid phase change. The latent heat of vaporization is the energy associated with liquid-vapour phase change.

11.3.1   Phase Equilibria for Water

Figure 11.1 shows the phase equilibrium diagram for water, and is broadly similar for most other substances with one important exception. Water is unusual in that it expands on freezing. The axes are pressure, volume and temperature so what you are seeing in effect is the surface represented by $f(p, v, T) = 0$, and clearly, the majority of it does not obey the ideal gas law.

Most noticeable is the parabolic looking curve in the middle of the diagram. In this region, water

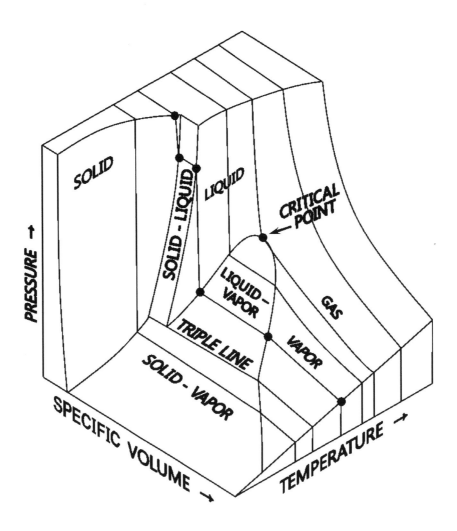

Figure 11.1: *P-v-T Phase Diagram for Water*

solid/liquid and/or vapour can coexist, the low volume leg of the 'parabola' is the 100% *saturated liquid line*, and the high volume leg is the 100% *saturated vapour line*. A constant temperature line between these two sides represents a change in volume at constant pressure of the vapour/liquid or vapour/solid mixture, ie a liquid/vapour mixture is evaporating/condensing depending which way the isotherm is traversed. One particular isotherm is the solid/liquid divisor, known as the *triple line*, below this there is a solid/vapour equilibrium, above it liquid/vapour equilibrium. On the triple line, solid, liquid and vapour are in phase equilibrium with each other. The peak of the 'parabola' is the *critical point* beyond which liquid and vapour states cannot be. Here and for volumes larger than the vapour line only vapour can exist, and is termed the *superheated vapour* region. This is a state where the temperature of the vapour is above the boiling point at the given pressure. For volumes smaller than the solid/liquid line only solid/liquid can exist, and this is termed the *subcooled* liquid/solid region.

In the saturated liquid-vapour region (inside the 'parabola') the mixture is termed *wet* (steam) and has a certain *dryness*. The dryness fraction the mass fraction of vapour present, $x = m_g/(m_g + m_f)$. For instance, if the mixture was on the saturated vapour line then the dryness fraction would be 1.

11.3.2 Psychrometry and Air Conditioning

Carnot Heat Pumps (section 3.7 on page 47) , with the Joule-Thompson effect operating on a Rankine type cycle (section 11.4, below) can effectively control temperature. However often we also wish to control the amount of water vapour in the air as well as this is a key metric in air quality conditioning for internal spaces in buildings. This requires an understanding of water vapour/air mixtures, and this area is known as psychrometry.

11.3.2.1 Dew Point

The amount of water vapour dry air can absorb is temperature dependent, the higher the temperature the more it can contain. The *dew point* occurs when the air is *saturated* with water vapour and it will start to condense. In many parts of the UK, the water vapour in the air condenses out during the night when the temperature drops, re-evaporating in the morning as the sun heats the air. Under normal conditions we treat the air/water vapour mixture as ideal and therefore Dalton's law (equation 11.2). The dew point defines the start of condensation of liquid water in a water vapour/air mixture and thus we should use the partial pressure of water vapour in the air to work out the dew point temperature. Therefore the dew point temperature is the saturation temperature at the given *partial* pressure of the water vapour.

11.3.2.2 Humidity Definitions

Humidity is a measure of the amount of mass of water vapour in the air, and we notice it by the 'stickiness' of the air. It has two common definitions. One is the specific (also known as the absolute) humidity, h_s, and is the mass of water vapour, m_v relative to the mass of dry air, m_a. The relative humidity, h_r is again the mass of water vapour in the air, but this time relative to the mass of water vapour in the air at saturated conditions (at that temperature), m_s. Recall that both of these gases are treated as ideal, and therefore $m = pV/RT$, where p is the partial pressure of the vapour.

$$h_a = \frac{m_v}{m_a} = \frac{p_v R_v}{p_a R_a} = 0.622 \frac{p_v}{p_a} \quad h_r = \frac{m_v}{m_s} = \frac{p_v}{p_s} \tag{11.4}$$

Relative Humidity is more usually quoted, and often as a percentage, since when the air approaches saturation, it feels 'sticky'. Remember also that humidity can change directly without adding or removing water vapour from the atmosphere, but by simply changing the temperature. Thus, by reducing the temperature of the air, we can approach saturated conditions and eventually reach the dew point as discussed in section 11.3.2.1.

11.4 Vapour Power Cycle

Previously, in section 3.9 on page 49 we considered heat engines where the working fluid was an ideal gas and obeyed the ideal gas equation and associated assumptions. An important class of heat engines rely on phase change for their heat transfer processes (for instance as shown in Figure 3.12 on page 53), and this has the important advantage that the heat transfers occur under near isothermal conditions. Coupled with the typically isentropic compression/expansion processes assumed for the work transfer stages heat engines involving phase change take on the character of the Carnot cycle, covered previously in section 3.8 on page 47.

The most common of these uses water as the working fluid, and is known as the Rankine Cycle. Closed cycle steam turbines are still the mainstay of large scale power generation systems. In the Rankine cycle the heat addition boils water, and heat rejection condenses it. In addition to the above mentioned advantage of near-isothermal heat transfer, others include (1) the vapour has a high energy density and (2) less compression work required to pressurise liquid rather than gas.

Theoretically a two phase Carnot cycle can operate, as shown by the cycle 1234 in figure 11.2 where process $2 \rightarrow 3$ represents heat addition and conversion of liquid water to vapour at constant temperature and process $4 \rightarrow 1$ the corresponding heat rejection and condensation process. In practice the compression and expansion processes are problematic with wet steam and the ideal Rankine cycle

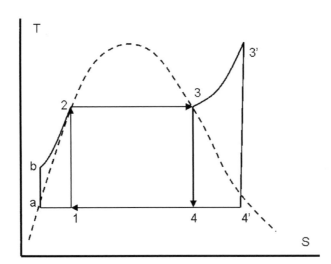

Figure 11.2: *The Two-phase Carnot and Ideal Rankine Cycle*

makes the following modifications.

- It is preferable to completely liquify the fluid before compression therefore in the Rankine cycle the fluid is condensed until the liquid line (point a on Figure 11.2). The liquid is then compressed to point b.
- Heat is then added from point b first to point 2, and then isothermally due to phase change to point 3, at which point the vapour line is reached.
- At this point more heat is added to take the maximum temperature up to 3', in the superheated vapour region.
- This allows the turbine stage $3 \rightarrow 4'$ to operate in the vapour only region, preventing damage to the blades.

Various modifications can be made the the Rankine cycle in order to improve the thermal efficiency.

- The condensor pressure can be lowered, reducing the heat rejection temperature (T_4).
- The boiler pressure can be increased, increased the boiling temperature (T_3).
- Increasing the superheat temperature $(T_{3'})$.
- Using reheat stages as discussed in section 3.13.2 on page 55.
- Use of combined cycles. A classic combination is to use a gas turbine (Brayton) cycle (section 3.13 on page 54) and a Rankine cycle together, where the heat rejection process of the gas turbine is the heat addition process of the Rankine cycle. This is feasible because the operating temperatures of the gas turbine cycle are significantly hotter than the Rankine cycle. Another combined cycle is to use the waste heat rejected for space heating, and this is common in large industrial sites, hospitals and so on.

12 | Heat Transfer

12.1 Introduction

Up until this point we have been interested in the effect a heat transfer *process* has upon a system, and its relationship to work transfers and internal energy changes of the fluid. This has been covered in sections 2.7.8, 2.9 and 3.2 on pages 33, 36 and 44 respectively. We have looked at heat transfer from control volume point of view, considering the Steady Flow Energy Equation in section 4.7 on page 63. We have defined and also looked at the balance of convective and diffusive flux through a 1-D fluid in section 2.6 on page 23 using thermal energy as an exemplar and examined the way in which heat flux through a duct wall affects the nature of a compressible fluid in section 10.7 on page 129. Finally, we have derived a general thermal energy conservation equation in section 8.7 on page 114 where we defined diffusion and convective flux terms, and also the reversible and irreversible work terms.

We know that heat transfer must occur across a temperature difference, and that this is an irreversible process, increasing the total entropy. We also know that flow convection can transport thermal energy, down a pipe for instance. We do not know however how convection can enhance the rate of heat transfer and we also do not know anything about radiative heat transfer - the 'warm glow' you feel when next to a fire.

12.2 Heat Transfer due to Conduction (revisited)

Heat conduction is a microscopic/molecular diffusional process as discussed in section 2.6.2 on page 23. The heat *flux* (heat flow per unit area) is defined in 1D in equation 2.12 and in 3D in section 8.4 on page 112.

12.3 Heat Transfer due to Convection

Heat transfer due to convection in this case is primarily related to transfer between a solid boundary and a fluid, and may be *free* or *forced*. Free convection is where natural buoyancy forces that arise due a solid/fluid heat transfer trigger fluid motion. Fluid motion in a kettle when it is heated is a good example. Forced convection occurs when the motion of the fluid is independent of the heating/cooling process, and water pumped through a car radiator is a good example of a forced convection heat transfer process. A sketch of a velocity and temperature profiles in the fluid due to a heated surface is shown in Fig. 12.1. Due to the no-slip boundary condition for the velocity at the wall the fluid velocity at the wall is zero and the heat transfer process *at* the wall is pure conduction. However the velocity near the wall will typically remove heat very efficiently and thus reduce the near wall fluid temperature. This increases the temperature gradient at the wall and thereby the total heat transferred from wall to fluid. Simple expressions for the combined effect of convection and diffusion are rare, and it is common for engineers to use a heat transfer coefficient, h, that acts as a 'lumped' parameter.

$$\dot{q} = h\left(T_s - T_\infty\right) \tag{12.1}$$

The state of the boundary layer, and whether it is laminar or turbulent greatly influences the heat flow, or, for a given temperature difference between the wall and the free stream fluid, greatly influences the heat transfer coefficient, h. A very useful non-dimensional number is the Nusselt Number, which defines the ratio of convective heat transfer relative to that due to conduction alone for a given length scale x

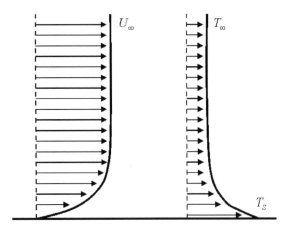

Figure 12.1: *Velocity (L) and Thermal (R) Boundary layers in Convective Heat Transfer*

$$Nu = \frac{hx}{k} \tag{12.2}$$

12.4   Heat Transfer due to Radiation

Heat (like light) can also be transferred by electromagnetic waves. The wavelengths relevant to thermal radiation are from the ultraviolet through the visible to the infrared scale, roughly 0.1 to $100\mu m$. When thermal radiation (like light) hits a material it can do one of three things;

- Be transmitted through it (like light through glass), with an efficiency ϵ_t.

- Be absorbed, and heat the material up (like placing a cup of water in a microwave), with an efficiency ϵ_a.

- Be emitted and heat up the surroundings that absorb the radiation, with an efficiency ϵ_e.

such that the sum of the efficiencies sum to unity, $\epsilon_t + \epsilon_a + \epsilon_e = 1$. Black bodies absorb *all* incident radiation and hence $\epsilon_a = 1$. Black bodies are also perfect emitters of radiation, and the radiation flux from a black body of temperature T is given as $\dot{q} = \sigma_{sb}T^4$, Stefan's Law. Emission from a real (non-black) body is termed grey body emission and is less that of a black body by a factor ϵ_g, which varies between 0 and 1.

Typically for solid materials $\epsilon_t = 0$ and at steady state the adsorption rate equals the emission rate for a body, in other words $\epsilon_a = \epsilon_e$. In this case, if a grey body at a temperature T is in an environment that acts as a black body at temperature T_b then the net rate of heat flux *from* the grey body will be.

$$\dot{q} = \epsilon_g \sigma_{sb}\left(T^4 - T_b^4\right) \tag{12.3}$$

12.5   Resistance Networks

A simple way to analyse conjugate heat transfer problems, for instance thermal conduction through layered materials, or heat transfer from a material that is radiating and undergoing convective heat transfer is to use a resistance network. This is is completely analogous with electrical resistance networks, we just have to define our 'thermal resistance' in an appropriate manner. For electrical (thermal) resistance, we have an electrical current (heat *flow* rate) that flows through it due to a

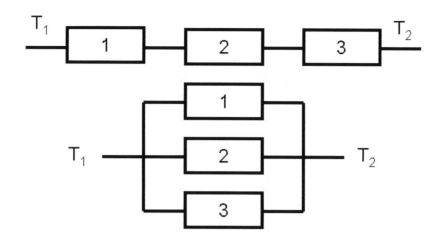

Figure 12.2: *Series and Parallel Resistance Networks between T_1 and T_2*

voltage (temperature) difference across it. Ohm's Law reflects this, $R = (V_1 - V_2)/I$. The thermal resistances for our three heat transfer processes are ;

- Diffusion (conduction) : $\dot{Q} = (kA/L)(T_1 - T_2)$, hence the diffusive thermal resistance is $R_D = L/kA$.

- Convection : $R_C = 1/hA$.

- Radiation : $R_R = 1/\left(\epsilon_g \sigma_{sb}\left(T_1^2 + T_b^2\right)(T_1 + T_b)\right)$.

The thermal resistances may then be arranged to suit the application to work out the total thermal resistance, and hence the net heat flux between the driving temperatures. Two base configurations are the series $R_T = R_1 + R_2 + R_3$ and parallel $1/R_T = 1/R_1 + 1/R_2 + 1/R_3$ networks as shown in Figure 12.2.

13 | Turbulence

13.1 Introduction

Turbulence is very important in engineering, and whilst quite easy to grasp a conceptual idea of what it is, how to account for its effects in engineering calculations is fraught with difficulties and dangers for the uninformed. This chapter is by no means even a brief introduction to the subject, but hopefully it will allow you to see the landscape, and point you where you want to go within it. Turbulence has already been introduced several times, quite appropriately, in chapter 5 which covered viscous flow. A few brief statements were made concerning the nature of turbulence in section 5.1 on page 75, and if you have forgotten them I strongly suggest looking again right now. This was necessary because we found that turbulent pipe flow causes a significant pressure drop (section 5.4.1 on page 78 when discussing the Moody diagram, and the sometimes effect of roughness). We also encountered turbulence again when considering developing boundary layers and the tendency for a boundary layer to separate in the presence of a adverse pressure gradient in section 5.7 on page 83. The common thread running through all of these introductory examples was that turbulence consumes energy, sometimes to our cost (pipe pressure drop) and sometimes to our benefit (delay of separation). This chapter formalises this energy consumption property and outlines the basic features of simple turbulent flow. We then find out why for most engineering systems turbulence is impossible to simulate directly and we must therefore employ some form of turbulence *model*. One industry standard method is briefly outlined, which serves to emphasise the degree of educated estimation that goes into its construction, and caution that should be employed upon use.

13.2 Velocity Fluctuations and the Kinetic Energy of Turbulence

Imagine we place an object in a wind tunnel and ensure the flow in the wake of the object is turbulent. We then place some sort of very small measurement device in the wake and measure the velocity in the flow direction $u(t)$ as a function of time t at a certain position, and we might see a trace like that shown in Fig. 13.1. The signal $u(t)$ looks random but does have some structure, and fluctuations about some steady value vary in both amplitude and also duration. What is being measured is a succession of turbulent eddies traversing the measurement device at more or less the average speed of the flow, most are fairly small and are only measured for a short duration, but occasionally, larger eddies appear. Although the velocity signal $u(t)$ is random in time the moments of the velocity signal are not, and this gives a little bit of information about the nature of the turbulence. The first moment of the signal is the average velocity $\overline{u}$, and is shown in Fig. 13.1. The difference between the mean and the actual velocity $u(t)$ at some time t is the velocity fluctuation $u'(t)$. Since at any point in time

$$u(t) = \overline{u} + u'(t) \tag{13.1}$$

and the average of $u(t)$ over many measurements is $\overline{u}$, then the average of the fluctuations is zero, $\overline{u'} = 0$. We can also work out the second moment of the signal, the mean of the square of the fluctuations about the mean (the variance), and this gives us directly the kinetic energy per unit mass of the fluctuations, $\overline{u'^2}$. Note this is not zero, because the square of the velocity fluctuations are always positive. So for our example, we have two kinetic energies, we have a kinetic energy per unit mass of the mean flow $\overline{u}^2/2$, just as we would in a laminar flow, and also a kinetic energy of the fluctuations about the mean flow, per unit mass, $\overline{u'^2}/2$. This latter quantity is the turbulent kinetic energy, $k/3$ (only a third of it because we have only considered one of the three velocity components) and defines how intense your turbulence is.

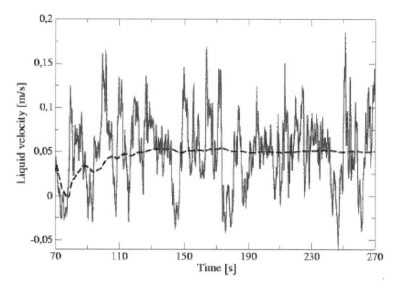

Figure 13.1: *The Fluctuating Velocity at a Point in Turbulent Wake of an Object*

13.3   The Energy Cascade

It was noted in section 5.1 on page 75 that turbulence is a dissipative process, it consumes energy. As long as the Reynolds Number of the flow is very large, turbulence can be thought of as an energy cascade process. Typically turbulent kinetic energy is generated by large eddies being shed off objects moving through the flow (or a flow moving past an object) so in that sense turbulence is generated at a large length scale l_0, the diameter of the object. These largest eddies also have a typical velocity fluctuation u_0 and hence a timescale $\tau_0 = l_o/u_0$. They have a turbulent kinetic energy therefore, of order u_0^2 and if we say they are being created at a frequency $1/\tau_0$ then the rate at which turbulent kinetic energy is being created at the large scales is $dk/dt = u_0^2/\tau_0 = u_0^3/l_0$. At steady state the rate at which turbulence energy is being created must equal the rate at which it is being destroyed (dissipated), and the rate of dissipation of turbulent kinetic energy is given the symbol ϵ, thus $dk/dt = -\epsilon$. What tends to happen is that turbulent energy is passed from large eddies (where it is created) to smaller eddies (where it is destroyed). It can be said (more or less) that energy *cascades* down the scale range.

This raises the question how small can the smallest eddies be, and it was also noted in section 5.1 on page 75 that the rate of dissipation ϵ is a viscous process (a function of the kinematic viscosity η then, purely on dimensional grounds, the smallest scales of turbulence can be defined. These are known as the Kolmogorov scales,

$$\eta \equiv (\nu^3/\epsilon)^{1/4} \quad , u_\eta \equiv (\epsilon\nu)^{1/4} \quad , \tau_\eta \equiv (\nu/\epsilon)^{1/2} \tag{13.2}$$

When we define a Reynolds Number based on the Kolmogorov scales, we find this is unity, suggesting viscous forces are very important at these scales and measurements confirm eddies of this order to be the smallest present in turbulent flow. A key feature of high Reynolds Number turbulence is that most of the turbulence energy is generated at the largest scales, and most of it dissipated at the smallest scales. In between, the 'inertial subrange' are larger eddies simply transferring turbulent kinetic energy to smaller eddies. Hence Richardsons little poem at the end of section 5.1 on page 75.

13.4   Length and Timescale Range : The need for Turbulence Models

The problem with typical engineering grade turbulence is there is simply too much information to handle on even the biggest computers. The ratio of the largest to the smallest scales, be they length, time or velocity can be expressed non-dimensionally in terms of the large scale Reynolds Number as follows,

$$\frac{\eta}{l_0} \equiv Re^{-3/4} \quad , \frac{u_\eta}{u_0} \equiv Re^{-1/4} \quad , \frac{\tau_\eta}{\tau_0} \equiv Re^{-1/2} \tag{13.3}$$

The length scale range ratio is not Re^3 so what is the problem ? Let us take a typical but not particularly large Reynolds Number, 10^4. The length scale ratio $l_0/\eta = 10^3$. This means the smallest eddy is 1000 times smaller than the largest eddy. So if we wanted to simulate turbulence in a 3D box one large eddy diameter on a side and resolve each small eddy with 10 control volumes then we would need 10^{12} control volumes. Likewise, for time, the ratio $\tau_0/\tau_\eta = 10^2$, in other words the largest eddy lasts 100 times longer than the smallest eddy. So if we wanted to simulate turbulence without any approximation, for say 10 large eddy lifetimes, and say we want to resolve the smallest eddy lifetime in 10 timesteps, then we would need to compute 1000 timesteps. Say we are really clever, and manage to reduce our computer code to n operations per control volume per time step, turbulence tells us that that is n $\times 10^{15}$ operations for our simulation, if we don't want to miss out on any physics. And note, as Re increases, these numbers increase astronomically. Try $Re = 10^6$, a typical cloud.

13.5 Reynolds Stresses

Luckily, as engineers, we are hardly ever interested in what *ALL* of the eddies are doing, and often we are not interested in *ANY* of them. Rather we are after the effect of turbulence on our current design problem, friction force in a pipe, mixing between two streams and so on. To be more precise, we want to know the effect of turbulence on the mean flow, because that will give us, say, the time averaged drag force. We can do this exactly, by substituting the mean and fluctuating parts of the instantaneous velocity (equation 13.1) into the instantaneous momentum conservation equation (eqn 8.7) and then average (in time here) again. The result is the *Reynolds Averaged* Navier-Stokes equations.

$$\frac{\partial}{\partial x_j}\left(\rho \overline{u_i}\,\overline{u_j}\right) = -\frac{\partial}{\partial x_j}\left(\rho\overline{u'_i u'_j}\right) - \frac{\partial}{\partial x_i}\left(\overline{p} + \frac{2}{3}\mu\frac{\partial \overline{u_k}}{\partial x_k}\right) + \frac{\partial \overline{\tau_{ij}}}{\partial x_j} + \rho g_i \tag{13.4}$$

This is the force-momentum equation for the mean (time averaged) specific momentum (velocity) and is exact. Notice an extra term has appeared, $\rho\overline{u'_i u'_j}$, and these are known as the Reynolds Stresses, and they arise from the decomposition and averaging of the non-linear convection term (2nd term of the rhs of equation 8.7). This term is real and significant in turbulent flow and to account for the effect of turbulence on the mean flow without calculating every single eddy we must approximate it somehow. The number of hours devoted to this problem by some of the finest engineering minds is vast.

The Reynolds Stress is a 2nd order symmetric tensor (the properties of which have been outlined in section 6.8 on page 101). The turbulent kinetic energy, k, introduced above in section 13.2 is twice the trace of the Reynolds Stress, such that $2k = \overline{u'_k u'_k} = \overline{u'_1 u'_1} + \overline{u'_2 u'_2} + \overline{u'_3 u'_3}$. When $i = j$ the Reynolds Stresses are termed the *normal* stresses, and thus $2k$ is the sum of the normal stresses. The turbulent normal stresses act a 'bit' like pressure. When $i /= j$ then the Reynolds Stresses are termed *shear* stresses. The turbulent normal stresses act a 'bit' like viscous shear stress.

Broadly there are two main methods to approximating turbulence, either by approximating some of the scales (filtering out the small ones, because these are where most of the effort is needed computationally) or all of them. Clearly a filtering operation is more accurate fundamentally because less information is lost, and these Large Eddy model methods are beginning to be used more extensively by industry. They still however require significant computational effort, and therefore simple and reliable turbulence models that do not require much more effort to solve than a laminar flow are still very much an industry requirement. Their exploration also allows some insight into the assumptions that underpin them, and thereby further insight into previous sections in this book.

13.6 Modelling the Reynolds Stresses : Eddy Viscosity Models

Recall the Newtonian laminar stress-strain relationship, first introduced in 1D in section 2.6.3 on page 24, equation 2.13 and also in 3D in tensor form in section 7.2 on page 106, equation 7.3. As discussed in section 2.6.2 on page 23, information (heat, momentum...) is transported by molecular collisions in *diffusion* processes. By analogy, Boussinesq (in 1877 no less) proposed that,

$$\rho \overline{u'_i u'_j} = \frac{2}{3}\delta_{ij}\rho k - \mu_t \left(\frac{\partial}{\partial x_j}\overline{u_i} + \frac{\partial}{\partial x_i}\overline{u_j} \right) \tag{13.5}$$

where μ_t is the turbulent eddy viscosity. It is a property of the *flow*, not the fluid. No *flow*, no turbulence, no turbulent viscosity. The analogy is that eddy collisions spread fluid information (heat, momentum ...) contained in these eddies in a '*diffusional*' spreading process in the same way molecular diffusion processes spread information in materials, still fluids and fluids in laminar flow. It is very important to note however that turbulent '*diffusion*' is totally a *convective* process, and is in general much more powerful than laminar diffusion processes. This is why people *stir* milk and sugar into their tea.

The advantage of equation 13.5 is that have we considerably simplified our problem, instead of needing to model 6 Reynolds Stress components in equation 13.4 we only have to model one turbulent viscosity. The kinematic form has dimensions $[L]^2[T]^{-1}$, or conversely a length $[L]$ and a velocity scale, $[V]$, so models are needed for these.

13.6.1 Prandtl's Mixing Length Model

Prandtl followed the above kinetic theory argument and proposed that the velocity scale should be a function of the typical fluctuating velocity, and a mixing length, as follows,

$$L = l_m \qquad V = l_m \frac{\partial}{\partial y}\overline{u} \ , \quad \mu_t = l_m{}^2 \frac{\partial}{\partial y}\overline{u}$$

He proposed that the velocity be a function of the mixing length and the (known) mean velocity gradient. Thus the model is not complete (we do not know the mixing length) and is also only valid for flows in one direction and spreading normal to that direction. This, and other more complicated models of this form all required a length scale assumption, which was only possible for very simple flow configurations. In this sense they were unclosed, and useless for general use. However they are more useful than it appears because common classical turbulent flows have simple spreading rates as outlined below.

13.6.1.1 Mixing Lengths for Prandtl's Mixing Length Model

For simple shear flows (wakes behind objects, round and planar jets emerging from orifici), turbulent spreading rates are extremely well behaved. For instance the spreading rate of a round turbulent jet is $l_m/\delta = 0.075$, where δ is the jet half width at some downstream location. Similar simple relations are well known for plane wakes, mixing layers and jets.

For walls, measurements suggest that the mixing length increases linearly away from the wall until a limit, where a constant mixing length applies. Defining δ as the edge of the boundary layer (see section 5.7.1 on page 83), in the linear region $0 \leq y/\delta \leq 0.25$ $l_m = \kappa y$. For $y/\delta \geq 0.25$, $l_m = 0.09\delta$. κ is known as Von Karman's constant and is about 0.4.

13.6.2 k-ϵ model

The k-ϵ model makes a key advance in that it is a closed model, and defines the turbulent viscosity in terms of the turbulent kinetic energy and the rate of its dissipation, and an empirical constant, C_μ.

$$\mu_t = \rho C_\mu \frac{k^2}{\epsilon} \tag{13.6}$$

Exact equations can be derived for both k and ϵ, though for the latter, this is of limited use. The $k - \epsilon$ model equations are,

$$\frac{\partial}{\partial x_j}\left(\rho \overline{u_j} k\right) = \frac{\partial}{\partial x_j}\left(\mu + \frac{\mu_t}{\sigma_k}\right)\frac{\partial}{\partial x_j}k - \rho \overline{u'_i u'_j}\frac{\partial}{\partial x_j}\overline{u_i} - \rho\epsilon \tag{13.7}$$

$$\frac{\partial}{\partial x_j}\left(\rho \overline{u_j}\epsilon\right) = \frac{\partial}{\partial x_j}\left(\mu + \frac{\mu_t}{\sigma_\epsilon}\right)\frac{\partial}{\partial x_j}\epsilon - C_{1\epsilon}\frac{\epsilon}{k}\rho \overline{u'_i u'_j}\frac{\partial}{\partial x_j}\overline{u_i} - C_{1\epsilon}\rho\frac{\epsilon^2}{k} \tag{13.8}$$

At first glance, we seem to have exchanged one constant l_m for five, $C_\mu, C_{1\epsilon}, C_{2\epsilon}, \sigma_k, \sigma_\epsilon$. However, the $k - \epsilon$ model is general in the sense that once the model constants are set, they will apply for all flows, be they simple shear flows like the smoke stack above, or complex three dimensional flows.

13.6.3 Turbulent Boundary Layers (revisited)

We have already encountered turbulent boundary layers, both internal (fully developed pipe flows, section 5.6 on page 81) and external (section 5.7 on page 83). This analysis applies primarily to the latter, but also qualitatively to the former, since it explains the dependence of pipe roughness on pipe Reynolds Number (section 5.6.1 on page 83).

If we consider a fully developed turbulent flow in the x_1 direction near a large plate normal to the x_2 direction, and ignoring gravitational forces, then equation 13.4 simplifies to

$$0 = \mu\frac{\partial^2}{\partial y^2}\overline{u} - \frac{\partial}{\partial y}\left(\rho\overline{u'v'}\right) - \frac{dp}{dx} \tag{13.9}$$

which implies the stress near the wall is constant, and is composed of two parts, a laminar, and a turbulent component. In other words,

$$u_\tau^2 = \frac{\tau_w}{\rho} = \frac{\mu}{\rho}\frac{\partial}{\partial y}\overline{u} - \overline{u'v'}. \tag{13.10}$$

u_τ is known as the friction or shear velocity and is used to define dimensionless wall velocities and lengths, including the non-dimensional wall normal length encountered in section 5.6 on page 81 as follows (note nomenclature change to (x, y) form)

$$u^+ \equiv \frac{u}{u_\tau} \quad , \quad y^+ \equiv \frac{\rho u_\tau y}{\mu} \tag{13.11}$$

Notice that y^+ is in reality a Reynolds number based on the wall normal distance. Equation 13.9 suggests two distinct regions exist, one where the viscous and the other where turbulence derived shear forces dominate.

13.6.3.1 Viscous Sub-layer

In this region $\mu(\partial u/\partial y) >> \rho\overline{u'v'}$ and thus $\mu(\partial u/\partial y) = \rho u_\tau^2$, which can be simplified to $u^+ = y^+$. This is simply a restatement of the laminar flow Newtonian stress-strain relationship. Typically, the viscous sub-layer has a thickness of $y^+ \leq 5$.

13.6.3.2 Log Law Layer

In this region $\mu(\partial\overline{u}/\partial y) << \rho\overline{u'v'}$ and the friction velocity is a function of the Reynolds Stresses only, $-\rho\overline{u'v'} = \rho u_\tau^2$. We have to approximate u_τ, and this is done by making the same assumptions as Prandtl in section 13.6.1.1 on page 147 where the mixing length is set to κy. Therefore $u_\tau = \kappa y(\partial u/\partial y)$ and the final non-dimensional relationship is known as the logarithmic *law of the wall*.

$$u^+ = \frac{1}{\kappa} ln(y^+) + C \tag{13.12}$$

The constants are usually taken to be $\kappa = 0.41$ and $C = 5.2$ for smooth walls. The log law is valid in the region $y^+ \geq 5$ and $y/\delta \leq 0.2$, where δ is the thickness of the boundary. For computational simulations, when making use of the log law assumption it is very important that the correct y^+ ranges are observed.

13.7 Summary and Further Reading

The presented turbulence models are a very small selection of a wide choice, chosen primarily to reinforce physical understanding of the assumptions made. Nevertheless, the $k - \epsilon$ model is still, even after several decades the industry standard. Numerous other two equation models exist, which have advantages in certain classes of flows, and one one equation model, the Spalart-Allmaras which is useful for aerodynamics applications. These two equation models offer a good compromise between ease of use and accuracy for the practising engineer. The next level up in complexity are Reynolds Stress Models, where transport equations for the 6 stresses are developed and solved. These models have clear advantages in certain situations but have not found widespread adoption, mainly due to their numerical instability. Large Eddy Simulation methods are completely different, and filter the Navier-Stokes equations to obtain a set of equations that define a spatially filtered velocity field, and instead adding a term to account for the small scale eddies that must be modelled. Again, in certain situations, LES models are a very attractive model choice.

Commercial computer codes for predicting turbulent flows offer all of these models, it is down to you to choose which is best. It should be remembered that turbulence is an important physical process for most engineering processes, and it is very easy, with the wrong model choice and/or set-up, that catastrophically wrong predictions can result. It is important to *understand* the implications of the model assumptions you are making by choosing one turbulence model over another.

The text by Pope is an excellent resource regarding turbulence and model choices.

14 | Computational Fluid Dynamics

14.1 Introduction

As shown in chapter 8 the governing equations of thermofluid flow are partial differential equations that in general do not have analytical solutions. Therefore we must solve them numerically and there are several different methods available. Here one method is presented which is known as the finite volume method. This is for two reasons, (1) the method directly solves the integral forms of the governing equations (e.g. equations 8.4, 8.7 and 8.9) and (2) it is the most common form of discretization used in commercial computational fluid dynamics (CFD) software packages. To introduce the finite volume method the example to be considered will be that of a single scalar quantity the evolution of which is described by the conservation equation in 2D non-vector differential form:

$$\frac{\partial \rho\phi}{\partial t} + \frac{\partial \rho u\phi}{\partial x} + \frac{\partial \rho v\phi}{\partial y} = \frac{\partial}{\partial x}\left(\Gamma_\phi \frac{\partial\phi}{\partial x}\right) + \frac{\partial}{\partial y}\left(\Gamma_\phi \frac{\partial\phi}{\partial y}\right) + \rho S_\phi \tag{14.1}$$

In the above ρ stands for density, U and V are the components of velocity in the x and y directions, Γ_ϕ is the diffusion coefficient for ϕ and S_ϕ represents the net formation rate of ϕ per unit volume. The diffusion coefficient Γ_ϕ can be expressed as $\Gamma_\phi \equiv \mu/\sigma_\phi$ where μ is viscosity and where σ_ϕ is the Prandtl or Schmidt number as appropriate. The quantity ϕ can stand for any transported scalar quantity such as internal energy, e_u or a velocity component. For the finite volume method we discretise and solve the integral form,

$$\int \frac{\partial \rho\phi}{\partial t}\, \partial V + \int \rho u\phi\, \partial S_x + \int \rho v\phi\, \partial S_y = \int \Gamma_\phi \frac{\partial\phi}{\partial x}\, \partial S_x + \int \Gamma_\phi \frac{\partial\phi}{\partial y}\, \partial S_y + \int \rho S_\phi\, \partial V.$$

The procedure which is to be followed is:

1. Overlay the solution domain with a computational grid.
2. At grid intersections (node points) store discrete values of the dependent variables.
3. Define control volumes, also called 'cells'.
4. Integrate the equation over a typical control volume, using Gauss' theorem.
5. Approximate the resulting integrals in terms of node values by, for example, presuming linear inter-nodal variations.
6. Analyse control volumes adjoining boundaries to allow application of boundary conditions.
7. Select a solution algorithm for solving the resulting set of algebraic equations.

The steps 1 to 6 are collectively called 'discretisation'. The first three relate to the solution domain whereas 4 and 5 refer to the differential equation. In general we must approximate three operations:

1. Integration (over a CV surface element)
2. Interpolation (of properties, stored at CV centres, to CV faces)
3. Differentiation (e.g. of the diffusion flux through a CV face)

In general all these operations should be implemented with numerical schemes of the same order to generate a complete scheme of said order.

14.2 Computational Grid, Control Volumes (CV) and Notation

In Figure 14.1 a domain has been chopped up into smaller volumes. The light solid lines are the computational *grid*, formed from coordinate surfaces, arbitrarily spaced. The outermost grid lines coincide with the surfaces of the domain. The grid intersections define *nodes* which are the locations at which the discrete values of ϕ will be calculated and stored. The outermost nodes lie on the intersections of the grid lines with the boundaries of the domain. Each node is enclosed in its own *control volume* or *cell*, delineated by the dashed lines, which are defined as lying midway between the grid lines. The outermost cell boundaries coincide with the boundaries of the solution domain, in this case the surfaces of the domain. We assume the mesh indexing system (i, j) is aligned with the coordinate system (x, y) and the control volume of interest is related to its neighbours in a structured way. The notation defines P as the CV of interest, N, S, E, W

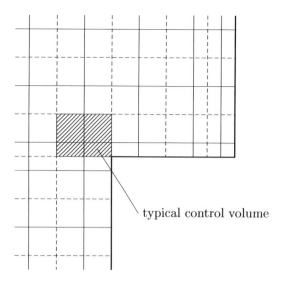

typical control volume

Figure 14.1: *A discretised domain showing a single control volume.*

neighbour CV's, n, s, e, w are the CV face centre locations in the N, E, S, W directions of the P CV and dx_p, dy_p the width and height of the CV of interest. For the collocated mesh described here all variables are stored at the cell centre location (P). Since uppercase (E, P) denote node position where information is stored, lowercase (e, w) denotes face locations where information must be interpolated to.

14.2.1 Interpolation Factors

Recall the conservation of mass equation (equation 8.4 on page 112), where we sum the mass fluxes over all the surface elements? We do the same in the finite volume method, and therefore we have to interpolate variables at (W, P, E) to the faces (w,e) to estimate fluxes. For uniform meshes this is easy e.g. $\phi_e = \frac{\phi_P + \phi_E}{2}$, $\phi_w = \frac{\phi_W + \phi_P}{2}$, but for non-uniform meshes, we require interpolation factors, e.g. $fx_P = fx_i = \frac{dx_P}{dx_P + dx_E} = \frac{dx_i}{dx_i + dx_{i+1}}$. Now we can define interpolation formulae for the east and west faces.

$$\phi_e = \phi_{i+1/2} = fx_P \phi_E + (1 - fx_P)\phi_P = fx_i \phi_{i+1} + (1 - fx_i)\phi_i$$

$$\phi_w = \phi_{i-1/2} = fx_W \phi_P + (1 - fx_W)\phi_W = fx_{i-1}\phi_i + (1 - fx_{i-1})\phi_{i-1}$$

Since this is based on linear interpolation, this is second order.

14.3 Discretization of the Diffusion Flux

As an example take the integral form of the conservation equation of a diffusion process is, $0 = \int \Gamma \frac{\partial}{\partial x_i} \phi \partial S_i$. The flux through the east face would be $J_e = \Gamma \frac{\partial}{\partial x} \phi \big|_e$ for instance. In discrete integral form for the 2D orthogonal mesh fluxes in the compass directions for the node P sum to zero, $0 = S_e J_e - S_w J_w + S_n J_n - S_s J_s$. In terms of the node values this would be $0 \approx C_e (\phi_E - \phi_P) - C_w (\phi_P - \phi_W) + C_n (\phi_N - \phi_P) - C_s (\phi_P - \phi_S)$ where the C terms involve geometry and the diffusion coefficient only. This can be recast as $A_P \phi_P = A_E \phi_E + A_W \phi_W + A_N \phi_N + A_S \phi_S$ where $A_P = A_E + A_W + A_N + A_S$. If we have a computational domain with I nodes in the x-direction and J nodes in the y-direction then each node (P say) is defined a location (i, j) in this matrix. In matrix form we may write $A\phi = 0$, and therefore to solve the 2D diffusion problem we have to solve this matrix equation. The task of finite volume discretization of the thermofluid conservation equations delegates to mapping flux coefficients to 'A' coefficients and ensuring the matrix is well conditioned (sparse, diagonally dominant). Considering now only the x-direction the discrete form of $\oint \Gamma \frac{\partial \phi}{\partial x} ds$, for uniform mesh spacing, gives

$0 = \Gamma \frac{\partial \phi}{\partial x}\Big|_e S_e - \Gamma \frac{\partial \phi}{\partial x}\Big|_w S_w$. Approximating the differential terms gives

$$\frac{\Gamma_e (\phi_{i+1} - \phi_i) S_e}{\frac{1}{2}(dx_{i+1} + dx_i)} - \frac{\Gamma_w (\phi_i - \phi_{i-1}) S_w}{\frac{1}{2}(dx_i + dx_{i-1})} \text{ or in compass notation } \Gamma_e \frac{(\phi_E - \phi_P) S_e}{\frac{1}{2}(dx_E - dx_P)} - \Gamma_w \frac{(\phi_P - \phi_W) S_w}{\frac{1}{2}(dx_P + dx_W)}.$$

We require matrix coefficients A_p, A_W and A_E such that $A_p\phi_p = A_E\phi_E + A_E\phi_W$ therefore

$$A_W = \frac{\Gamma_w S_w}{\frac{1}{2}(dx_p + dx_W)}, \quad A_E = \frac{\Gamma_e S_e}{\frac{1}{2}(dx_E + dx_p)}.$$

Traditionally, the terms are grouped such that $D_e = \frac{\Gamma_e}{\frac{1}{2}(dx_E + dx_P)} S_e$, therefore $A_W = D_w$, $A_E = D_e$ and $A_p = A_E + A_W$.

14.4   Discretization of the Convective Flux

The 1D steady convection equation is $\frac{\partial}{\partial x}(\rho u \phi) = 0$ and in integral form $\int \rho u \phi \, dS = 0$. In discrete integral form $\rho u \phi S|_e - \rho u \phi S|_w = 0$. Again it is traditional to define a convective flux term (in reality the mass flow through the CV face) $F_e = \rho u S|_e$, $F_w = \rho u S|_w$. For uniform meshes approximate the face values as $\phi_e = \frac{\phi_P + \phi_E}{2}$, $\phi_w = \frac{\phi_P + \phi_W}{2}$ and so our discrete convective flux equation becomes

$$\frac{F_e}{2}\phi_p + \frac{F_e}{2}\phi_E - \frac{F_w}{2}\phi_p - \frac{F_w}{2}\phi_W = 0.$$

This presents a problem, because we need to use a matrix structure of the form $(A_E + A_w)\phi_p = A_p\phi_p = A_E\phi_E + A_W\phi_W$. To solve this problem we take the continuity equation, $\frac{\partial}{\partial x}(\rho u) = 0$. In integral form $\oint \rho u \, dS = 0$ and in discrete integral form, $\rho u S|_e - \rho u S|_w = 0$ or $F_e - F_w = 0$. We multiply this by ϕ_p, $F_e\phi_p - F_w\phi_p = 0$ and subtract it from our scalar transport equation to give

$$-\frac{F_e}{2}\phi_p + \frac{F_w}{2}\phi_p = -\frac{F_e}{2}\phi_E + \frac{F_w}{2}\phi_W.$$

Now, for the general matrix equation $A_p\phi_p = A_E\phi_E + A_w\phi_W$, $A_E = -\frac{F_e}{2}$, $A_W = \frac{F_w}{2}$ ensuring $A_P = A_E + A_W$. For non-uniform meshes we can employ interpolation factors to give $A_E = -fx_P F_e$ and $A_W = (1 - fx_W) F_w$.

14.5   Discretization in Time

We can consider time dependence by integrating

$$\rho_P V_P \frac{d\phi_P}{dt} = A_E (\phi_E - \phi_P) + A_W (\phi_W - \phi_P) + A_N (\phi_N - \phi_P) + A_S (\phi_S - \phi_P) + Q_P$$

over the time interval t_n to $t_{n+1} \equiv t_n + dt$, where dt is the 'time step'. There are a variety of ways in which this can be accomplished. We are going to cover the two extremes, implicit methods and explicit methods. We can also choose the order of method, and here we only consider 1st order schemes. To illustrate these we will first write the above equation as $\frac{d\phi_P}{dt} = f(x,t)$ where $f = f(\phi_E, \phi_W, \phi_N, \phi_S, \phi_P, x, t)$.

14.5.1   Explicit Time Discretization

A first order forward time difference approximation yields:

$$\frac{\phi_P^{n+1} - \phi_P^n}{dt} = f(\phi(x_i)_n)$$

With this approximation, we have:

$$\rho_P V_P \frac{\left(\phi_P^{n+1} - \phi_P^n\right)}{dt} = A_E \left(\phi_E^n - \phi_P^n\right) + A_W \left(\phi_W^n - \phi_P^n\right) + A_N \left(\phi_N^n - \phi_P^n\right) + A_S \left(\phi_S^n - \phi_P^n\right) + Q_P.$$

It can be seen that each ϕ_P^{n+1} is independent of its neighbouring ϕ^n and can therefore be obtained by straightforward substitution of the ϕ^n on the right-hand side. A suitable (very simple) algorithm based on the above would therefore involve the following steps; starting from prescribed initial grid ϕ and advancing a time increment, e.g.

1. Assemble the coefficients A_α, P_P, and Q_P etc.
2. Systematically scan the grid and compute the ϕ_P^{n+1}.
3. Advance another increment dt and repeat from 1, after setting ϕ_P^n equal to the just calculated ϕ_P^{n+1}.

This is a simple and apparently economical procedure. However a simple analysis shows that stable solutions can only be obtained only under certain conditions. For the case of uniform properties and a square mesh with $dx = dy$ the following constraints must be satisfied:

$$\frac{\Gamma dt}{dx^2} \leq \frac{1}{4}, \quad \frac{|v|\, dy}{\Gamma} = \mathrm{Pe}_V \leq 2, \quad \frac{|u|\, dx}{\Gamma} = \mathrm{Pe}_U \leq 2.$$

The constraint on the maximum value on the 'diffusion number' $\frac{\Gamma dt}{dx^2}$ is equivalent to requiring that during the computational time interval dt the ϕ 'wave' should not propagate more than a cell dimension dx. These are severe constraints. For problems involving diffusion only, if dx is reduced to improve spatial accuracy, then $dt_{\max} \propto dx^2$ and the time step must be reduced more than proportionally. If convection is involved then the mesh must be sufficiently fine to ensure that the cell Peclet numbers are less than two everywhere. This is prohibitively expensive for the majority of flows. Note these constraints are *not* imposed by accuracy considerations, they apply even when ϕ is uniform. Usually acceptable accuracy can be obtained with values of dt considerably larger than $dt_{\max}$ and with cell Peclet numbers much greater than two so there is a considerable cost penalty. The algorithm becomes *unstable* when the constraints are not observed and the net outcome is that the explicit algorithm is usually *not* cost-effective.

14.5.2 Implicit Time Discretization

A first order backward time difference approximation yields:

$$\frac{\phi_P^{n+1} - \phi_P^n}{dt} = f\left(\phi(x_i)_{n+1}\right)$$

and now

$$\rho_P V_P \frac{\left(\phi_P^{n+1} - \phi_P^n\right)}{dt} = A_E \left(\phi_E^{n+1} - \phi_P^{n+1}\right) + A_W \left(\phi_W^{n+1} - \phi_P^{n+1}\right) + A_N \left(\phi_N^{n+1} - \phi_P^{n+1}\right) +$$
$$A_S \left(\phi_S^{n+1} - \phi_P^{n+1}\right) + Q_P.$$

Both the left and the right hand sides have new time level information, ϕ^{n+1}, so a set of algebraic equations for each 'P' must be solved simultaneously. The implicit form can be written, after rearrangement, as:

$$A_P \phi_P^{n+1} = \sum_{\alpha=N,S,E,W} A_\alpha \phi_\alpha^{n+1} + B_P$$

where

$$A_P = \sum_{\alpha=N,S,E,W} A_\alpha + \frac{\rho_P \phi_P}{dt} + P_P \phi_P \quad \text{and} \quad B_P = Q_P \phi_p + \frac{\rho_P \phi_P}{dt} \phi_P^n.$$

The advantage of the (first order) implicit method is that it is unconditionally stable and thus time steps of any size may be used. In the limit of very large time steps the steady state discrete matrix is recovered. This disadvantage is the solution of one node (at P) is dependent on its neighbours, and therefore separate equations for each node must all be solved together.

14.6   Discretisation of Source Term

A 1D steady, integrated convection-diffusion equation, with sources is of the form $\oint \rho u\phi \, dS = \oint \Gamma \frac{\partial \phi}{\partial x} \, \partial S + S$ and the general matrix equation is of the form

$$(A_p - P)\,\phi_p = A_E \phi_E + A_w \phi_w + Q$$

where $A_E = D_e - fx_P F_e$, $A_W = D_w + (1 - fx_W)\,F_w$ and the source term has been linearise and made a function of ϕ_p, $S = Q + P\phi_p$. To increase the diagonal dominance of the matrix, and thus the stability of the numerical method $P < 0$. Often however, numerical discretization of source terms is more art than science.

14.7   Discretization of Boundary Conditions

The sources Q, P are also used to correct the matrix coefficients at the domain boundaries to allow boundary conditions to be defined in the matrix.

14.7.1   Constant value diffusion flux (wall) boundary

Figure 14.2 shows the 'w' face of the P CV defined as a wall where $\phi_W = \phi_w = \phi_{wall}$ (taken to be a constant, temperature say). At the wall the convective and diffusive fluxes are $F_w = F_{wall} = 0$ and $D_{wall} = \frac{\Gamma_{wall}}{dx_{wall}} S_{wall}$, respectively. For the node P, assuming central differencing for convection $F_e(fx_P \phi_E + (1 - fx_P)\phi_P) - 0 = D_e(\phi_E - \phi_P) - D_{wall}(\phi_P - \phi_{wall})$ where $A_E = D_e - fx_P F_e$, $A_w = 0$, $Q = D_{wall}\phi_{wall}$ and $P = -D_{wall}$. Q and P are the same for 'e' boundaries also. Again $P < 0$ always.

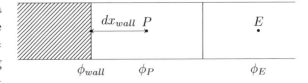

Figure 14.2: 'w' wall boundary.

14.7.2   Constant flux diffusion flux (wall) boundary

Now imagine instead of a constant value at our wall we have a constant flux. J_{wall} has units of ϕ/m^2s, $+ve$ into CV so $J_{wall} = const. = D_{wall}(\phi_P - \phi_{wall})$. Therefore ϕ_{wall} must be calculated from J and ϕ_P, $\phi_{wall} = \frac{J}{D_{wall}} + \phi_P$. Then as before $A_E = D_e - fx_P F_e$, $A_w = 0$, $Q = D_{wall}\phi_{wall}$ and $P = -D_{wall}$. Obviously is the wall flux is zero (if the variable is energy, the wall is adiabatic, then $A_W = Q = P = 0$.

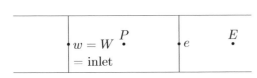

Figure 14.3: 'w' inlet boundary.

Q, P are the same for the 'e' boundary also.

14.7.3   Prescribed Flow (inlet) boundary

For simplicity assume zero diffusion flux, then for the flow boundary on the 'w' face as shown in Figure 14.3 then $\phi_w = \phi_W = \phi_{inlet}$. The convective flux for node P in discrete integral form is $\rho u\phi S|_e - \rho u\phi S|_{inlet} = 0$ giving the discrete form

$$\frac{F_e}{2}\phi_P + \frac{F_e}{2}\phi_E - F_w \phi_{inlet} = 0$$

assuming a uniform mesh. Continuity multiplied by ϕ_p gives $F_e\phi_p - F_w\phi_p = 0$, subtracting gives

$$-\frac{F_e}{2}\phi_P + \frac{F_e}{2}\phi_E - F_w\phi_{inlet} + F_w\phi_P = 0.$$

Since we need to use a matrix structure of the form $(A_p - P)\phi_p = A_E\phi_E + A_w\phi_w + Q$ $A_E = -\frac{F_e}{2}$, $A_W = 0$, $P = -F_w$ and $Q = F_w\phi_{inlet}$.

14.8 Outflow boundary

Generally we do not know very much about the flow conditions at the outlet boundary so the general practice is to assume zero gradients normal to the boundary, for instance on the 'e' face as shown in Figure 14.4. Assuming zero diffusion flux and zero gradient along a streamline normal to the outflow boundary face gives the convective flux for node P in discrete integral form is $\rho u\phi S|_e - \rho u\phi S|_{wt} = 0$. At the boundary $\phi_e = \phi_P$ therefore we have

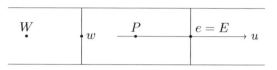

Figure 14.4: 'e' outflow boundary.

$$F_e\phi_P - \frac{F_w}{2}\phi_P - \frac{F_w}{2}\phi_W = 0.$$

Following the usual steps with the continuity equation gives

$$-\frac{F_w}{2}\phi_W + \frac{F_w}{2}\phi_P = 0$$

for node P. Since we need to use a matrix structure of the form $(A_p - P)\phi_p = A_E\phi_E + A_w\phi_w + Q$ therefore $A_E = 0$, $A_W = \frac{F_w}{2}$, $P = 0$ and $Q = 0$.

14.9 Boundedness of Discretization Schemes

The solutions of partial differential equations like equation 14.1 describing the evolution of a strictly conserved scalar variable with constant properties are such that the maximum and minimum values must lie on the domain boundaries. To show this consider the situation where a local maximum occurs at time $t = 0$. At this maximum the gradients $\partial\phi/\partial x$ and $\partial\phi/\partial y$ will be zero and $\partial^2\phi/\partial x^2 < 0$ and $\partial^2\phi/\partial y^2 < 0$. Therefore at the maximum $\partial\phi/\partial t$ will be negative, and ϕ must decay, since $\phi(t + dt) = \phi(t) + \partial\phi/\partial t\, dt$. Likewise, where a local minimum occurs $\partial^2\phi/\partial x^2 > 0$ and $\partial^2\phi/\partial y^2 > 0$, $\partial\phi/\partial t > 0$ will be positive and ϕ at the minimum will increase. Thus, ϕ is *bounded* and for steady solutions, $\min(\phi_B) \leq \phi(\mathbf{x}) \leq \max(\phi_B)$ the value of the variable in the domain never lies outside that on the boundaries.

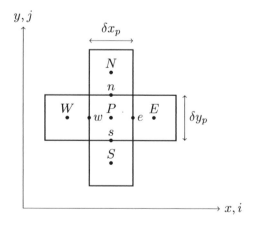

Figure 14.5: A set of 5 control volumes showing the compass notation, and upper case nodes and lower case locations.

The principle outlined above may be viewed as a generalisation of the second law of thermodynamics. If ϕ represents temperature then satisfaction of the principle implies that heat transfer occurs only from high temperatures to low temperatures. This is obvious for pure diffusion problems, but it also applies equally well to systems with convection and diffusion. Satisfaction of the above boundedness criteria by the solutions of the discrete finite volume equations is highly desirable and often essential. The steady flow form of the finite volume approximation to a *strictly* conserved scalar quantity – a quantity for which there are no sources or sinks:

$$A_P\phi_P = A_E\phi_E + A_W\phi_W + A_N\phi_N + A_S\phi_S \quad \text{so} \quad \phi_P = \frac{A_E\phi_E + A_W\phi_W + A_N\phi_N + A_S\phi_S}{A_P}.$$

Recall $A_P = A_S + A_N + A_W + A_W$ so ϕ_P is a weighted average of the ϕ's at the surrounding node points. Clearly if ϕ_P to satisfy the boundedness criteria then the 'weights' must be positive and this requires that the coefficients, $A_S, A_N, A_W, A_E \geq 0$. Note this is not equivalent to defining a numerically solvable system. Consider the steady form of equation 14.1 with no source term, and we discretise this on a 2D orthogonal mesh as shown in Figure 14.5 using central differencing for both convective a diffusive fluxes. It is a simple yet 2nd order scheme, so this seems sensible. Our influence coefficients become Assuming central differencing on a uniform mesh for both the diffusive and the convective fluxes then

$$A_N = \left(\frac{\Gamma_n}{\frac{1}{2}(dx_N + dx_P)} - \frac{\rho_n V_n}{2}\right) S_n$$

and similarly for the other compass directions. Using the D and F notation, $A_N = D_n - F_n/2$ and since $A_N > 0$ for a bounded scheme, $D_n - F_n/2 \geq 0$, or $1 - \frac{F_n}{2D_n} \geq 0$, or $F_n/D_n = \text{Pe}_n \leq 2$. For this to be the case then the cell Peclet numbers must be less than 2. For $\text{Pe}_n \geq 2$, then we may expect unbounded solutions to arise, as is shown in the following example. The model example is as introduced in section 2.6.5.4 on page 28, namely $u\frac{d\phi}{dx} = \Gamma\frac{d^2\phi}{dx^2}$ where $\Gamma \geq 0$, $U \geq 0$, Γ and U constant, BC's: $\phi(0) = 0$; $\phi(1) = 1$. The analytic solution to this problem is:

$$\phi = \frac{1 - e^{\frac{xu}{\Gamma}}}{1 - e^{\frac{u}{\Gamma}}}.$$

Consider now the following mesh, where $\phi_W = 0$ and $\phi_E = 1$. The standard central difference finite volume approximation to the problem gives

$$\frac{U}{2\Delta x}(\phi_E - \phi_W) = \frac{\Gamma}{\Delta x^2}(\phi_E - 2\phi_P + \phi_W).$$

Solving for P

$$\phi_P = \phi_E\left(\frac{1}{2} - \frac{\text{Pe}}{4}\right) + \phi_W\left(\frac{1}{2} + \frac{\text{Pe}}{4}\right)$$

where $\text{Pe} \equiv \frac{u\Delta x}{\Gamma}$. Thus when $\text{Pe} = 0$, $\phi_P = \frac{\phi_E + \phi_W}{2}$, when $\text{Pe} = 2$, $\phi_P = \phi_W$ and as $\text{Pe} \to \infty$, $\phi_P \to -\infty$ since $\phi_W \leq \phi \leq \phi_E$.

14.9.1 Bounded Convective Schemes

So far to evaluate the cell face values associated with the convection terms a linear inter-nodal variation has been presumed. If instead the cell face value is presumed to be that prevailing at the upstream node then this is referred to as 1st order upwind differencing.

$$(\rho u\phi)_e \approx (\rho_e u_e)\phi_e \quad : \quad u_e > 0 \quad \phi_e = \phi_P \quad u_e \leq 0 \quad \phi_e = \phi_E$$
$$(\rho u\phi)_w \approx (\rho_w u_w)\phi_w \quad : \quad u_w > 0 \quad \phi_w = \phi_W \quad u_w \leq 0 \quad \phi_w = \phi_P.$$

When this is applied to the 1D model problem above this gives:

$$\frac{u}{\Delta x}(\phi_P - \phi_W) + \left(\frac{U\Delta x}{2}\frac{d^2\phi}{dx^2}\right) = \frac{\Gamma}{\Delta x^2}(\phi_E - 2\phi_P + \phi_W).$$

Solving for P gives

$$\phi_P\left(1 + \frac{2}{\text{Pe}}\right) = \phi_E\left(\frac{1}{\text{Pe}}\right) + \phi_W\left(1 + \frac{1}{\text{Pe}}\right),$$

hence ϕ_P is bounded for all values of Peclet number. Therefore this method guaruntees to give bounded numerical solutions and does promote stable numerical solutions. However it is not very accurate, it tends smear gradients and acts to increase the apparent viscosity of the fluid to such an extent that it should only be used as a method of last resort. Various higher order convection schemes are available but most are locally not bounded producing instability in numerical solutions and numerical artefacts in the final solutions. One class, total variation diminishing (TVD) methods are however bounded and where possible, 2nd order accurate.

14.10 Matrix Solution

Whatever the character of the equation to be solved, a matrix is always generated. The matrix form of the entire equation set is $\mathbf{A}\phi = \mathbf{b}$ where ϕ is the vector of N unknowns (N is the number of grid nodes in the mesh), $\mathbf{A}$ is an $N \cdot N$ matrix (in theory) containing the coefficients A_P and A_α and $\mathbf{b}$ is the vector of explicit source terms. Direct methods such as Gaussian elimination place minimum constraints on the coefficient matrix but are not cost-effective when, as is usually the case in finite volume analysis, the matrix contains many zero entries and has a regular-banded structure, features which the alternative methods exploit. Iterative methods such as point Gauss-Seidel iteration exploit the matrix regularities outlined above, at the price of some loosely-defined constraints on coefficients. In particular, a *sufficient* condition for the applicability of these methods is the *diagonal dominance* requirement, which states that at each grid node, $|A_P| \geq \sum_\alpha |A_\alpha|$ where the inequality must hold at least one node. The definitions of the coefficients outlined ensure that the above requirement is satisfied.

14.10.1 Convergence Criteria

It is necessary to determine when iterative solution schemes have 'converged'. There are no definitive criteria for iteration convergence. A commonly used criterion is one where the entire array is checked for: $\max \left| \phi_P^{(n+1)} - \phi_P^{(n)} \right| \leq \epsilon_1$ or $\max \left| \left(\phi_P^{(n+1)} - \phi_P^{(n)} \right) / \phi_r \right| \leq \epsilon_2$. This is a necessary but not sufficient condition, because it can be satisfied when convergence is slow. A better practice is to supplement the above by test on the '*residual sources*' R of the finite volume equations. This is defined as: $R_P^{(n)} = A_N \phi_N^{(n)} + A_S \phi_S^{(n)} + A_E \phi_E^{(n)} + A_W \phi_W^{(n)} + B_P - A_P \phi_P^{(n)}$. Evidently $R_P^{(n)}$ is simply a measure of the degree to which the local 'conservation balance' is satisfied by the $\phi^{(n)}$ field. A suitable normalised measure for the whole mesh is: $\sum \left| R_\alpha^{(n)} / Q \right| \leq \epsilon_3$ where the summation is over all interior nodes, Q is a reference flux of ϕ and where ϵ_3 is a 'small' number, typically 10^{-3}.

14.10.2 Under-Relaxation

In this practice, which can be applied to any iterative procedure, the solution at each iteration, denoted by $\phi^{(n)}$, is regarded as provisional, and modified according to: $\phi_P^{n+1} = r \phi_P^{n+1^*} + (1-r) \phi_p^n$ where r is the relaxation factor the value of which must be specified. If $0 \leq r \leq 1$ then the practise is referred to as '*under-relaxation*'. This can sometimes help convergence, particularly for non-linear problems and/or if the steady flow forms of the equations are being solved, i.e. if $dt \to \infty$. In the case where the steady flow equations are being solved under-relaxation is introduced implicitly by modification of the coefficients. In the case of steady flow the finite volume equations become

$$A_P \phi_P^{n+1} = r \left(\sum_{\alpha=N,S,E,W} A_\alpha \phi_\alpha^{n+1} + B_P \right) + (1-r) A_P \phi_P^n.$$

In a code the coefficients are modified: $B_P' \equiv B_P + \frac{1-r}{r} A_p \phi_P^n$ and $A_P' = \frac{A_P}{r}$.

14.11   Calculation of Pressure

Since we are already discretising and solving the momentum equation there is no 'direct' equation for pressure and in implicit "pressure correction" methods, a pressure field is guessed and corrected until mass and momentum conservation is achieved. Here the "pressure equation" is the mass conservation equation. Starting from the momentum equation (equation 8.7 on page 8.7) the algebraic form (units N) is:

$$A_p^u u_{pi} = \sum_{n=E,N,T,W,S,B} A_n^u u_{ni} + S_i - \frac{\partial P}{\partial x_i} \partial V \tag{14.2}$$

where $\sum A_n u_{ni}$ contains all implicit convention and diffusion terms. Writing equation 14.2 in terms of u, (denoted u_i^f),

$$u_i^f = \frac{S_i + \sum A_n^u u_{ni}^f}{A_p^n} - \frac{\partial V}{A_p^u} \frac{\partial P}{\partial x_i}.$$

Note that u_i^f conserves momentum, but not mass. We set the first term of the RHS to u_i^m such that

$$u_i^m = \frac{S_i + \sum A_n^u u_{fni}^f}{A_p^u} \text{ therefore } u_i^f = u_i^m - \frac{\partial V}{A_p^u} \frac{\partial P}{\partial x_i}.$$

We define u_i^c as the velocity field from the continuity equation thus u_i^c obeys mass, but not momentum conservation. To conserve mass *and* momentum, we correct u_i^f by u_i' to match U_i^c, i.e.

$$U_i^c = U_i^f + U_i' = U_i^m - \frac{\partial V}{A_p^u} \frac{\partial P'}{\partial x_i}.$$

Note we are applying a pressure *correction*, P'. Substituting the new expression for U_i^c back into the continuity equation gives (a diffusion equation for P')

$$\frac{\partial \rho}{\partial t} + \frac{\partial}{\partial x_i} \left\{ \rho u_i^m - \rho \frac{\partial V}{A_p^u} \frac{\partial P'}{\partial x_i} \right\} = 0 \text{ or } \frac{\partial \rho}{\partial t} = \frac{\partial}{\partial x_i} \left\{ \frac{\rho \partial V}{A_p^u} \frac{\partial P'}{\partial x_i} \right\} - \frac{\partial}{\partial x_i} (\rho u_i^m).$$

If the fluid is a gas (follows gas law) we can decompose the density, $\rho = \rho^o + \rho' = \rho^o + \frac{p'}{RT}$ and create a convection term by decomposing density into a correction term.

$$\frac{\partial}{\partial x_i} \rho u_i^m \Rightarrow \frac{\partial}{\partial x_i} (\rho^o u_i^m) + \frac{\partial}{\partial x_i} \left(\frac{u_i^m}{RT} P' \right).$$

We then decompose the time term into an implicit and an explicit part, $\rho^n = \rho^o + \rho' = \rho^o + \frac{\partial \rho}{\partial P} \Big|_T P' = \rho^o + \frac{P'}{RT}$ so

$$\frac{\partial}{\partial t} \rho = \frac{\rho^n - \rho^o}{\Delta t} = \frac{1}{\Delta t} \frac{P'}{RT}$$

and the final conservation equation defining the pressure correction is

$$\left(\frac{P'}{RT\Delta t} \right) + \frac{\partial}{\partial x_i} \left(\frac{u_i^m}{RT} P' \right) = \frac{\partial}{\partial x_i} \left\{ \frac{\rho^n \partial V}{A_p^u} \frac{\partial P'}{\partial x_i} \right\} - \frac{d}{dx_i} (\rho^o u_i^m).$$

Note that the ρ^n has not been decomposed, for this, the first term on the RHS becomes

$$\frac{\partial}{\partial x_i} \left\{ \frac{\rho^n \partial V}{A_p^u} \frac{\partial P'}{\partial x_i} \right\} = \frac{\partial}{\partial x_i} \left\{ \frac{\rho^o \partial V}{A_p^u} \frac{\partial P'}{\partial x_i} \right\} + \frac{\partial}{\partial x_i} \left\{ \frac{P' \partial V}{RT A_p^u} \frac{\partial P'}{\partial x_i} \right\}.$$

The overall solution procedure is

1 Obtain new estimates for U_i^f, and also work out the diffusion coefficient in the P' equation.
2 Obtain a new estimate for P'.
3 Correct the velocities (conserving mass) using the pressure correction gradient field.
4 Correct the pressure field using the pressure correction field.
5 Go round again.

14.11.1 Pressure Decoupling

For a uniform mesh, the pressure gradient is

$$\int_{w}^{e} -\frac{\partial P}{\partial x}\, dV = -\left(P_e - P_w\right)S = -\left(\frac{(P_E + P_P)}{2} - \frac{(P_P + P_W)}{2}\right)S = -\left(\frac{P_E - P_W}{2}\right)S.$$

In other words, the pressure gradient at P does not depend on the pressure at P. This permits pressure checkerboarding to arise. If all the even nodes had pressure $+P$ and all the odd nodes had pressure $-P$ using central differencing the pressure gradient at every node would be zero. A remedy is to stagger the momentum CV's, relative to the scalar (pressure) CV. As shown in Figure 14.6 a Scalar (pressure) CV has (I, J) indexing, U CV has (i, J) indexing, V CV has (I, j) indexing. Now, discretizing the pressure gradient over U and V CV's provides for the U velocity

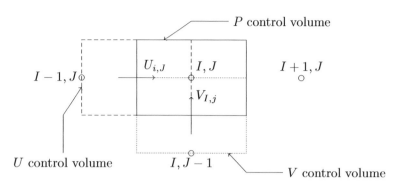

Figure 14.6: *Staggered mesh nomenclature.*

$$\int_{w}^{e} -\frac{\partial P}{\partial x}\, dV_u = -\left.\left(P_e - P_w\right)\right|_u S = -\left.\left(P_P - P_W\right)\right|_P S = -\left.\left(P_{I,J} - P_{I-1,J}\right)\right|_P S.$$

In algebraic form, the U and V momentum equations (f superscript denotes field conserving momentum) are

$$A_{i,J} u_{i,J}^{f} = \sum_{n=E,N,W,S} A_n u_n^{f} + Q_U - \left(P_{I,J} - P_{I-1,J}\right) S_Y,$$

$$A_{I,j} v_{I,j}^{f} = \sum_{n=E,N,W,S} A_n v_n^{f} + Q_V - \left(P_{I,J} - P_{I,J-1}\right) S_X$$

Advantages of staggered mesh methods include no interpolation of pressure to velocity CV faces, no interpolation of velocity to pressure (and all other scalar) CV faces and exact treatment of velocity on domain boundaries. Disadvantages include 3 meshes required for 2D problems, 4 for 3D, interpolation of transport coefficients to velocity faces. Finally the method only really works for orthogonal meshes. General pressure-velocity solution on non-orthogonal meshes uses co-located meshes and a pressure gradient smoothing method attributed to Rhie and Chow.

14.12 Summary and Further Reading

A basic introduction to CFD, using the finite volume method in 2D for orthogonal meshes is presented. The method can be easily generalised to 3D non-orthogonal meshes, as used in most commercial CFD codes, a good text for further reading is by Peric.

A | Bibliography

There is nothing new in this book, it is simply an attempt to deliver the core knowledge, without the window dressing and at a significantly lower price tag. Nevertheless this knowledge has come from a variety of sources over the years and the following books are good for the following things in my view.

A.1 Maths Texts

- 'Flatland: A Romance of Many Dimensions' by Edwin A. Abbott , 1923 : An amusing if slightly misogynistic view of life in two dimensions. Free on Project Gutenburg.
- 'Engineering Mathematics' by Kreyszig : A solid and easy going text. Advanced version also very good.
- 'Vectors, Tensors and the Basic Equations of Fluid Mechanics', Rutherford Aris : All you ever wished you didn't have to know about tensor operations. Very cheap.

A.2 Thermofluids Texts

- 'Thermodynamics', Cengel and Boles
- 'Engineering Fluid Mechanics' Crowe, Elger and Roberson

These are two good texts for Thermodynamics and Fluid Mechanics. the Cengel and Boles text is very good, and will have all the information most engineering students will need. The Crowe et al book is a very good introductory text for fluids. The 9th edition is concise and is the last edition Clayton Crowe contributed to, and I think it is better than the 10th edition, which is getting fat and becoming just another fluids textbook.

A.3 Numerical Methods Texts

- 'Numerical Methods in engineering and science', Graham de Vahl Davis : A clear introduction to solving equations numerically
- 'Computational Methods for Fluid Dynamics', Ferzinger and Peric : Methods for solving thermofluid partial differential equations

A.4 More Advanced Undergraduate/Graduate Texts

- 'Fundamental Mechanics of Fluids', IG Currie
- 'Transport Phenomena', Bird Stewart and Lightfoot
- 'Turbulent Flows', SB Pope
- 'Boundary-Layer Theory' - H Schlichting

These, with a copy of Cengel and Boles will ensure you have virtually all of what you need. A copy of Perry's Chemical Engineers Handbook is also useful for referencing information.

4